CELL BIOLOGY RESEARCH PROGRESS

ONCOPROTEINS: TYPES AND DETECTION

CELL BIOLOGY RESEARCH PROGRESS

Additional books in this series can be found on Nova's website under the Series tab.

Additional E-books in this series can be found on Nova's website under the E-book tab.

CANCER ETIOLOGY, DIAGNOSIS AND TREATMENTS

Additional books in this series can be found on Nova's website under the Series tab.

Additional E-books in this series can be found on Nova's website under the E-book tab.

CELL BIOLOGY RESEARCH PROGRESS

ONCOPROTEINS: TYPES AND DETECTION

JEREMY R. DAVIS
EDITOR

Nova Biomedical Books
New York

Library of Congress Cataloging-in-Publication Data

Oncoproteins : types and detection / editor, Jeremy R. Davis.
p. ; cm.
Includes bibliographical references and index.
ISBN 978-1-61761-551-1 (hardcover)
1. Tumor proteins. I. Davis, Jeremy R.
[DNLM: 1. Oncogene Proteins. QZ 202]
RC268.42.O555 2010
616.99'407--dc22

2010029838

Published by Nova Science Publishers, Inc. † New York

Contents

Preface

Oncoproteins are any proteins coded by an oncogene and they play an important role in the regulation or synthesis of proteins linked to tumorigenic cell growth. Some oncoproteins are accepted and used as tumor markers. This book presents topical research data in the study of oncoproteins, including the cellular redox environment and oncongenes/tumor suppressors in breast cancer cells; cell surface oncoproteomics; site specific references for oncoprotein tumor markers; oncoproteins in tropical infections; and the E7 oncoprotein of human papillomavirus.

Chapter 1 - This review is intended to provide a broad overview of interplay between changes in the cellular redox environment and oncogenes/tumor suppressors in breast cancer cells. To survive, malignant cells must adapt to a variety of environmental conditions, including persistent oxidative stress and genomic instability. Tumor cells promote their survival capacity by shifting their redox environment to more reductive conditions and increasing the expression of metallothioneins, Bcl-2, as well as through activation of redox sensitive prosurvival pathways including phosphatidylinositol 3-kinase (PI3K)-Akt and MAPK. Changes in the intracellular redox homeostasis alter the activity of tumors suppressors leading to their inactivation or complete loss, which ultimately stimulates tumor cells proliferation and invasiveness. Adaptation of tumor cells to the changed environment necessitates changes in gene expression regulated by redox sensitive transcription factors such as HIF, AP-1, p53, etc. This review also examines interactions between prosurvival signaling pathways and ROS generating NADPH oxidases.

Chapter 2 - The c-Myb transcription factor has been implicated in the control of proliferation and differentiation in many different cell types. While several mechanisms for regulating it have been identified, evidence suggests that the activity of c-Myb can be easily modified and varies in different cell types, and even in a time-dependent manner. For example, c-Myb binds and regulates the promoters of genes that control different parts of the cell cycle, such as CDC2, CCNB1 and CCNA1, suggesting that the activities and specificities of c-Myb, and which genes it regulates, vary during the cell cycle. Wild type c-Myb and its truncated and mutated oncogenic derivative v-Myb regulate distinct sets of genes, suggesting that mutations in c-Myb, which mimic the effects of post-translational modifications, completely change its activity and direct it to different target genes. These findings suggest that protein-protein interactions, subject to the effects of mutations and post-translational modifications, are the predominant modifiers of c-Myb activity, leading to dynamic changes in its specificity in real time. Here we will address the mechanisms that regulate the changes in c-Myb activity in different cell types and during the cell cycle with an emphasis on how these mechanisms go awry in transformed cells.

Chapter 3 - Tumour invasion and metastasis are often responsible for a high mortality rate in cancer patients. The spread of cancer involves multiple signalling pathways mediated by cell surface proteins and generally results in the loss of cell-cell adhesion leading to cell migration and invasion of distant tissues. Many transformed cells may be eliminated by immune surveillance, but others evade detection. Cell surface profiling is of special interest in oncoproteomics as it may provide a more accurate and reliable diagnosis and prognosis and identify new therapeutic targets. Such profiles may provide disease signatures that correlate with disease subtypes, reflecting the mutated genetic program of malignant cells. Extensive immunophenotypes of live-cells captured on antibody microarrays provide a quick, reproducible scan of the cell surface for patterns of known tumour biomarkers. Discovery proteomics projects, including glycoprotein-targeted approaches, provide potential biomarkers that can be screened on large clinical cohorts using antibody microarrays. The application of these technologies is discussed in the context of leukaemia and metastatic melanoma with their potential use in the clinic.

Chapter 4 - There are several presently used oncoprotein tumor markers. These markers are very useful in oncology and widely used worldwide. An important problem in interpretation of tumor marker result is the reference. The site specific references are needed for better interpretation of tumor maker result. In this specific paper, the author will present the approach on finding for new site specific references for oncoprotein tumor markers. An example from Thai context will be used as a model.

Chapter 5 - Epstein-Barr virus (EBV) is a ubiquitous herpesvirus. Epstein-Barr virus (EBV)-encoded latent membrane protein 1 (LMP1) expression is closely associated with the malignant transformation of undifferentiated nasophayngeal carcinoma (NPC). LMP1 plays a strong role in activating the anti-apoptotic response in the host cells. It could also stimulate growth and division of the infected nasopharyngeal epithelial cells. Further, LMP1 expression is associated with the aggressiveness of the disease. Recent studies revealed that LMP1 could alter the host's epigenetic machinery and activate expression of oncogenic microRNAs. Thus, LMP1 is considered as the key mediator in transforming normal nasopharyngeal epithelial cells. In terms of regulation, expression of LMP1 is regulated largely by epigenetic mechanisms such as CpG island methylation. However, the key triggering factors for LMP1 upregulation is not yet found. Recent study indicated that the infected EBV is involved in controlling LMP1 levels. EBV-encoded microRNA could bind to the 3' UTR of the LMP1 transcript and hence hindering the translation of LMP1 protein. Understanding the regulation dynamics of LMP1 is essential in developing novel prevention or treatment regime for the disease.

Chapter 6 - Generally, oncoprotein is mainly used in the oncological investigation and study. However, the non oncological aspect of oncoprotein is of interest. In this specific article, the author will discusses on the aspect of oncoprotein in tropical infection. This is a really forgotten aspect of oncoprotein in tropical medicine. The scenario in several tropical infections including malaria will be presented in this brief article.

Chapter 7 - Generally, laboratory investigation for oncoprotein is a new set of laboratory investigation in oncology. The investigation plays very important roles in diagnosis and following up of cancerous patients. Similar to any investigation in laboratory medicine, the problem of laboratory investigation can be expected. The problem of false positive in laboratory investigation for oncoprotein is an interesting topic in medical oncology. Since it can lead to diagnostic pitfall, the concern on this problem in required. In this specific article, the author hereby briefly discuss on this specific topic.

Chapter 8 - Human Papillomaviruses (HPVs) are involved in the etiology of at least 10% of all human cancers and are mainly implicated in the development of pre-malignant and malignant lesions of the ano-genital tract. In particular, genital High-Risk HPVs are responsible for nearly 100% of cases of cervical cancer, which represents the second most prevalent cancer in women worldwide.

The carcinogenic risk associated with HPV infection is primarily due to the activity of two viral oncoproteins, E6 and E7. Indeed, their expression is necessary for the maintenance of the malignant phenotype, and is always retained in HPV-positive cancer cells.

Although malignant transformation results from the synergistic and complementary effects of E6 and E7, the latter is reported to be the main oncoprotein of HPVs. One of the major aims of this review is to update the understanding of the role of E7 in HPV-associated carcinogenesis, focusing on the binding with its cellular targets and the effects induced by these associations. The interaction of E7 with its main target, the tumor suppressor protein retinoblastoma (pRb), will be described, with emphasis on the biological consequences and the domains involved. However, as E7 oncogenic properties are not only attributable to this association, the significance of other relevant interactions will be evaluated as well.

Due to the essential role of E7 in cervical carcinogenesis, the lack of homology with cellular proteins and the exclusive expression in cancer cells, this oncoprotein represents the main target for cervical cancer therapy. The therapeutic potential of strategies aimed at E7 inhibition for the treatment of HPV-associated lesions will be addressed. Among others, the use of small interfering RNA (siRNA), intracellular antibodies ("intrabodies"), ribozymes, anti-sense oligonucleotides and aptamers to knock out E7 at the gene/protein level will be covered. The results of these studies will be reviewed with the aim of highlighting E7 inhibition as a promising approach for the treatment of HPV-associated cancer.

Chapter 9 - Human Papillomaviruses (HPVs) are involved in the etiology of at least 10% of all human cancers and are mainly implicated in the development of pre-malignant and malignant lesions of the ano-genital tract. In particular, genital High-Risk HPVs are responsible for nearly 100% of cases of cervical cancer, which represents the second most prevalent cancer in women worldwide.

The carcinogenic risk associated with HPV infection is primarily due to the activity of two viral oncoproteins, E6 and E7. Indeed, their expression is necessary for the maintenance of the malignant phenotype, and is always retained in HPV-positive cancer cells.

Although malignant transformation results from the synergistic and complementary effects of E6 and E7, the latter is reported to be the main oncoprotein of HPVs. One of the major aims of this review is to update the understanding of the role of E7 in HPV-associated carcinogenesis, focusing on the binding with its cellular targets and the effects induced by these associations. The interaction of E7 with its main target, the tumor suppressor protein retinoblastoma (pRb), will be described, with emphasis on the biological consequences and the domains involved. However, as E7 oncogenic properties are not only attributable to this association, the significance of other relevant interactions will be evaluated as well.

Due to the essential role of E7 in cervical carcinogenesis, the lack of homology with cellular proteins and the exclusive expression in cancer cells, this oncoprotein represents the main target for cervical cancer therapy. The therapeutic potential of strategies aimed at E7 inhibition for the treatment of HPV-associated lesions will be addressed. Among others, the use of small interfering RNA (siRNA), intracellular antibodies ("intrabodies"), ribozymes, anti-sense oligonucleotides and aptamers to knock out E7 at the gene/protein level will be covered. The results of these studies will be reviewed with the aim of highlighting E7 inhibition as a promising approach for the treatment of HPV-associated cancer.

In: Oncoproteins: Types and Detection
Editor: Jeremy R. Davis

ISBN: 978-1-61761-551-1

Chapter 1

Redox State and Oncogenes in Breast Cancer

E.A. Ostrakhovitch
Department of Chemistry, University of Western Ontario,
London, Ontario, Canada

Abstract

This review is intended to provide a broad overview of interplay between changes in the cellular redox environment and oncogenes/tumor suppressors in breast cancer cells. To survive, malignant cells must adapt to a variety of environmental conditions, including persistent oxidative stress and genomic instability. Tumor cells promote their survival capacity by shifting their redox environment to more reductive conditions and increasing the expression of metallothioneins, Bcl-2, as well as through activation of redox sensitive prosurvival pathways including phosphatidylinositol 3-kinase (PI3K)-Akt and MAPK. Changes in the intracellular redox homeostasis alter the activity of tumors suppressors leading to their inactivation or complete loss, which ultimately stimulates tumor cells proliferation and invasiveness. Adaptation of tumor cells to the changed environment necessitates changes in gene expression regulated by redox sensitive transcription factors such as HIF, AP-1, p53, etc. This review also examines interactions between prosurvival signaling pathways and ROS generating NADPH oxidases.

Introduction

Breast cancer still remains the second cause of death worldwide regardless of advances in novel therapeutic agents mostly due to development of metastasis rather than tumor burden [97, 33, 49]. Initiation and progression of breast cancer involve accumulation of genetic and epigenetic changes that results in activation of oncogenes and inactivation of tumor suppressors. These changes account for limitless replicative potential and evasion of apoptosis, thus providing growth and survival advantages to the malignant cells. Changes in intracellular redox environment leading to accumulation of oxidizing reactive oxygen species (ROS) result in accumulation of mutations, deletions, gene amplification and rearrangements that initiate activation of several proto-oncogenes, such as Ras and ErbB2, and inactivation of some tumor suppressor genes [139]. Regulation of DNA repair is sensitive to cellular redox state and can be modulated by estrogens, which are also involved in indirect modulation of redox potential [154, 3]. Hyperactivation of oncogenes and mutations in tumor suppressor genes induce unchecked cell division and tumor growth via substantial changes in thiol reactivity and redox sensitive signaling. Thus, activation of ErbB2 (Neu), an oncogenic receptor tyrosine kinase, initiates a disruption of mammary epithelial architecture at the apical–lateral border, thus contributing to tumor progression [59, 241, 248, 206].

Oncogenes

Oncogenes are perturbed normal cellular genes, which have been shown to confer properties to tumor cells or tumor-like cells essential for their behavior as tumor cells. They are mostly genes that control crucial physiological functions related to cell growth and differentiation. There are the following well known oncogenes in breast cancer: ErbB2/Her2, Ras, c-fos, c-jun, C-myc, cyclins D1 and E, PI3-kinase, Akt, EIF-4E, BRCA1 and BRCA2. In tumor cells, changes associated with accumulation of successive mutations, which involve activation of oncogenes and suppression of tumor suppressor genes, lead to neoplastic transformation characterized by uncontrolled proliferation and dedifferentiation of progeny cells. It was suggested that there are, in fact, only few genes that can actually promote cancer formation in humans when mutated. Among these genes are BRCA, PIK3CA (which encodes the p110 catalytic subunit of phosphatidylinositol-3-OH kinase), and Akt1 and TP53 genes, which stand at the top of the list of most frequently mutated genes in various cancers including breast cancer [175, 173, 207, 32, 198, 202]. The gene coding for p110α, PIK3CA, is mutated in human cancers, which carry a single or one of three "hot spot" mutations [197, 198]. Mutation of a catalytic subunit of phosphatidylinositol-3-OH kinase upregulates the lipid kinase activity of the mutant protein leading to enhanced phosphorylation of Akt, S6K, 4EBP and GSK3β [91, 198], which stimulates cell transformation, induces anchorage-independent growth and causes tumors in animal model systems [10, 91, 95].

The keys players in regulation of redox homeostasis

The reduction-oxidation (redox) state of the cell is defined by the balance between the levels of oxidizing and reducing equivalents or, in other words, by the balance between cellular antioxidant defense system and prooxidants. The intracellular redox environment is fundamental to cell homeostasis and physiological function. Changes in the reduced/oxidized ratio dynamically control the intracellular redox state and, at the same time, represent a fundamental regulatory mechanism in living cells to control the activities of enzyme and gene expression implicated in cellular processes such as growth, differentiation, and death. Reduction or oxidation of protein redox-active groups is a key event in enzymatic reactions

[222]. However, imbalance in redox regulation leads to development of various pathological conditions including cancer.

Within the cell, the redox status depends on the relative amounts of the oxidized and reduced form of specific redox couples such as NAD^+/NADH, $NADP^+$/NADPH, GSSH/GSH (glutathione), TrxSS/Trx$(SH)_2$ (thioredoxin), cystine/cysteine, metallothionein/thionein, ascorbate/dehydroascorbate, oxygen/superoxide, H_2O/H_2O_2 and others. Accumulation of oxidants including reactive oxygen species (ROS) shifts the intracellular redox environment to highly oxidized state and stimulates oxidation of various molecules. Strong oxidants are kept under tight control by anti-oxidant proteins, such as superoxide dismutase (SOD), catalase, glutathione peroxidise (GPx) and glutathione reductase (GR), and by non-enzymatic compounds ascorbate and glutathione [56, 81]. Redox molecules such as glutathione represent the most prevalent non protein intracellular thiol, which is considered a major thiol-disulfide redox system with the highest intracellular concentration among other redox active ones [48]. Glutathione exists as a reduced predominant form (GSH), as a disulfide form (GSSG), or as mixed disulfide (GSSR) with protein thiols [142]. The GSSG/GSH ratio reflects the redox status since the oxidation of a limited amount of GSH can drastically change this ratio and thereby greatly affect the redox status within the cell. In contrast to other redox systems, absolute concentrations of GSH and GSSH have an impact on the reduction potential of this redox pair [199]. The major source of reducing equivalents for the GSH and thioredoxin systems is NADPH generated by the pentose phosphate pathway, since $NADP^+$/NADPH couple has the highest reduction potential of -374 mV. Thioredoxin, peroxiredoxins, and metallothioneins are another important examples of redox couples that maintain intracellular redox environment.

An increase in cellular glutathione levels is frequently observed in human cancers and is closely associated with increased resistance to chemotherapy and radiotherapy [193, 147]. The concentration of thiols was higher in the tissues of cancer patients than in normal tissues, with its highest level at stage III of the disease [194]. An increase in the GSH level and alteration of its intracellular distribution in cancer cells correlate with overexpression of anti-apoptotic Bcl-2 [226, 76]. Breast carcinomas show higher activities of glutathione reductase and catalase than the normal mammary tissue. Low molecular weight sulfur containing proteins including sulfiredoxin involved in maintenance of cellular redox homeostasis were also linked with oncogenic transformation. Overexpression of sulfiredoxin triggers an alteration in the expression and phosphorylation of cell cycle regulators $p21^{CIP1/WAF1}$, $p27^{Kip1}$ and p53, stabilizes the phosphatase PTEN and activates phosphatase PTP1B [124].

Thioredoxins

Thioredoxins (Trx) are a class of low molecular weight redox proteins that undergo reversible reduction-oxidation of two active-site cysteine residues with the reduction catalyzed by NADPH-dependent flavoenzyme thioredoxin reductase. The redox activity is essential for the growth stimulating function of Trx; therefore, redox inactive forms of Trx lacking active site Cys32 and/or Cys35 loose their activity. Thioredoxins and thioredoxin reductase, the expression and localization of which positively correlate with those of Trx, form a powerful network involved in regulation of cell proliferation, redox regulation of gene expression, expression of redox sensitive transcription factors such as p53 and NFκB and anti-apoptotic functions by maintaining a reducing intracellular environment [130, 221, 166, 220]. The PTEN gene is often inactivated at late stages in sporadic breast cancers via hypermethylation of PTEN promoter, which contributes to the progression, invasiveness and metastasis [106, 195]. Thioredoxin 1 (Trx-1) directly binds to apoptosis signal-regulating kinase 1 (ASK1) and tumor suppressor PTEN to inhibit their activity [35, 131]. Furthermore, an increased association between PTEN and thioredoxin was detected in tumors in which activities of thioredoxin (Trx-1) and thioredoxin reductase (TrxR1) were significantly elevated [114].

Thioredoxin-1 (Trx-1), a redox oncoprotein, is highly expressed in a number of human primary tumors, including breast cancer, thyroid, prostate and colorectal carcinoma, and malignant melanoma, and its expression associates with tumor aggressiveness, invasiveness, and decreased patient survival [70, 20, 104, 162, 130, 129, 34, 220]. Numerous studies showed that thioredoxin overexpression was associated with facilitated proliferation and increased resistance to chemotherapeutic agents in vitro [163, 11, 66]. Recombinant human thioredoxin added to culture medium without serum stimulated proliferation of breast cancer MCF-7 and UA2087 cell lines, whereas neither redox inactive nor mutant human thioredoxin, C32S/C35S, were able to stimulate proliferation, colony formation, and tumor formation [69, 67, 66]. Increased nuclear Trx-1 accumulation in breast cancer cells was related to cisplatin-resistance [38]. Although thioredoxin knockdown did not significantly affect the basal level of epithelial breast cancer MCF-7 cell death, it did enhance cisplatin induced cell cycle arrest and cell death [155]. Similar, upregulation of thioredoxin in primary breast cancer patients was to blame for their lower response to docetaxel as compared to patients with low thioredoxin expression [108]. The expression of thioredoxin and peroxiredoxin1 was also elevated in tamoxifen-resistant MCF-7 cells, in which the intracellular level of peroxide was significantly decreased as compared with parental cells [109]. Upregulation of these proteins was controlled by NF-E2-related factor2 (Nrf2)/ARE since tamoxifen resistance was partially reversed by Nrf2 siRNA.

However, pro-oxidant and pro-apoptotic role of Trx was also demonstrated in anthracycline-mediated apoptosis in epithelial breast cancer cells [184]. Treatment of MCF-7 cells overexpressing thioredoxin with daunomycin (or daunorubicin), an anthracycline derived antibiotic, stimulated apoptosis via enhanced NADPH-oxidase mediated generation of superoxide and increased p53 activity [184]. Furthermore, it was shown that overexpression of Trx-1 promotes apoptosis via activation of caspase 8, while its down-regulation or expression of mutant redox-inactive form results in decreased expression of caspase-7 and p53, suppression of apoptosis and suppression of superoxide generation in response to antracycline treatment. [134].

Thioredoxin interacting protein (TXNIP), which negatively regulates TRX function and is also known as thioredoxin binding protein 2 (TBP-2), was implicated in the suppression of tumor development and metastasis [232, 167, 5]. Thioredoxin binding protein 2 expression is reduced in human primary breast and colon tumors compared with adjacent tissue [30]. Direct transcriptional repression of TBP-2 promoter by cyclin-dependent kinase inhibitor $p21^{CIP1/WAF1}$ induces Trx secretion and angiogenesis in epithelial breast cancer MCF7 cells [111]. Cells transfected with TBP-2 siRNA promoted cell invasion and blocked the anti-angiogenic effect of $p21^{CIP1/WAF1}$ siRNA. However, overexpression of thioredoxin binding protein-2 in these cells suppresses tumor cells growth [167].

Peroxiredoxins

All members of peroxiredoxins (Prxs) family which are divided into six groups (Prxs I-VI) can reduce H_2O_2 using electrons from thioredoxin (Trx) or other substances. Similarly to thioredoxin-1, peroxiredoxin-1 is highly overexpressed in breast cancer and its expression level associated with the tumor grade, since overexpression of Trx-1 leads to upregulation of peroxiredoxin [21, 34, 168, 103]. The levels of messenger RNA (mRNA) for both Prx I and Trx-1 in normal human breast tissue were very low compared to other major human tissues, whereas their levels in breast cancer were the highest among other solid cancers (colon, kidney, liver, lung, ovary, prostate, and thyroid) [34]. Peroxiredoxins exist in low and high molecular weight forms and their cellular functions depend on their oligomeric form. It was suggested that the disulfide form of 2-cys peroxiredoxins favors a dimeric structure, whereas oxidized sulfenic form of 2-cys prefers a decameric structure [237, 239]. Low molecular weight Prxs 1 and 2 show peroxidase activity, whereas high molecular forms gain chaperon functions [96]. At low hydrogen peroxide concentrations, Prxs 1 and 2 are mainly present in low molecular weight forms and act as highly specific reducing agents, while in the presence of high concentrations of hydrogen peroxide, overoxidation of the peroxidatic cysteine (Cys51) results in conversion of Prxs into high molecular forms that acquire chaperon function and protect cells from apoptosis. For some time it had been assumed that the catalytic efficiency of Prx is significantly lower than that of glutathione peroxidase and catalase. However, later it was shown that catalytic activity of Prx 2 is similar to that of catalase and glutathione peroxidase (rate constant of $1.3x10^7$ $m^{-1}s^{-1}$) [172].

Prolonged inactivation or lack of Prx 2 results in increased H_2O_2 production, activation of PDGFR and consequently increased cell proliferation and migration. Prx 1 binds to the Src homology 3 (SH3) domain of c-Abl, a cell cycle protein, inhibiting its tyrosine kinase activity [85]. The oxidized oligomeric states of Prx 1 and 2 were observed during mitogenic signaling and oxidative stress [176]. The presence of oxidized oligomeric forms of Prxs associates with suppression of cyclin D1 expression and cell cycle arrest, whereas in the presence of reduced Prx, cell cycle progression resumes. Thus, the Prx oligomeric state is controlled by the redox state of the cell and Prx conversion into oxidized high molecular weight forms regulates cyclin D1 expression and cell cycle progression [110]. However, Prx itself controls the redox state by keeping the H_2O_2 level low under resting conditions while permitting increased production of H_2O_2 during signal transduction as a result of oxidation of Prx key cysteine residues to cysteine sulfinic or sulfonic acid, leading to Prx inactivation [238]. Inactivation of Prxs provides conditions required for triggering signaling cascades of redox-regulated transcription factors. Indeed, mammary epithelial cells lacking Prx 1 produce higher level of reactive oxygen species and show substantial activation of Akt associated with increased PTEN oxidation [113, 31]. The binding of peroxidase Prx1 to PTEN promotes PTEN tumor suppressive function by stabilizing PTEN binding to the plasma membrane and therefore protecting the phosphatase activity under oxidative stress during transformation. Upregulation of Prxs 1 and 2 leads to more invasive phenotype of transformed cancerous cells also characterized by upregulation of Akt and loss of PTEN function [43, 34, 103, 168].

Redox environment in regulation of oncogenes

ErbB2

The product of the ErbB2 (Her2/neu in human) proto-oncogene is the second member of the human epidermal growth factor receptor family of tyrosine kinase receptors. It has been suggested to associate with tumor aggressiveness, prognosis and responsiveness to hormonal and cytotoxic agents in breast cancer patients [112]. The ErbB proteins are a family of receptor kinases that include four closely related members: epidermal growth factor receptor (EGFR/ErbB1), ErbB2/neu, ErbB3, and ErbB4. However, of all members of this family, ErbB2 plays a pivotal role in ErbB-mediated activities in breast cancer. The intracellular signalling pathways regulated by Her2 involve Ras- mitogen-activated protein kinase (MAPK), MAPK-independent S6 kinase and phospholipase C-gamma networks. Overexpression of the receptor tyrosine kinase erbB2 (Her2 in humans) in human mammary carcinomas generally associates with a poor prognosis in breast and ovarian cancers. ErbB2/Her2 is constitutively activated in approximately 80% of all breast cancers and in particular in drug resistant metastatic breast cancers. Positivity for c-ErbB2 oncoprotein occurs mainly in large cell ductal and infiltrating ductal breast carcinomas [210]. In cells expressing high level of ErbB2, growth factor is inhibitory, and such inhibition associates with erbB2 phoshorylation. Autoactivated ErbB2 activates Akt and thus causeselevated expression of Akt-regulated pro-survival genes [228, 191]. Cells with blocked ErbB2 displayed a dose-dependent increase in ROS production and cell death, and these processes were reversed by an antioxidant, N-acetylcysteine [73].

Estrogen receptor

Exposure to estrogens increases the risk of breast cancer and estrogens have been implicated to be carcinogens. It was proposed that the estrogen receptors (ER) may contribute to estrogen carcinogenesis by transduction of the hormonal signal with subsequent formation of reactive oxygen species and/or through a receptor-mediated pathway, which may also involve indirect modulation of the intracellular redox state. Accumulation of genotoxic estrogen metabolites in the nucleus enhances DNA damage via modulation of intracellular redox environment by shifting it into a more oxidized state. Thus, treatment of estrogen receptor alpha (ER-α)-positive MCF-7 cells and ER-alpha-negative MDA-MB-231 cells with 10 nM 17beta-estradiol (E2) resulted in a marked decrease in the ability of MCF-7 cells to metabolize peroxide, which associated with a decrease in catalase activity and total glutathione levels, while these effects were not observed in the MDA-MB-231 cells [154]. However, E2 enhanced the activities of glutathione peroxidase and superoxide dismutases in MCF-7 cells. Furthermore, metabolism of estrogens to catechols and further oxidation to redox active/electrophilic o-quinones correlate with genotoxicity of estrogens [230]. The endogenous estrogen quinones primarily form unstable N3-adenine or N7-guanine DNA adducts, consequently converting it in mutagenic apurinic sites, whereas equine estrogen quinones formed from estrogens generate a variety of DNA lesions, including apurinic sites, DNA strand cleavage and oxidation of DNA bases [24]. The highly redox-active catechol also may target redox-sensitive enzymes involved in stress response cascades implicated in regulation of cell proliferation, leading to breast epithelial cells transformation into a tumorigenic phenotype. Furthermore, thioredoxin (Trx) and Trx reductase (TrxR) together associate with the DNA-bound estrogen receptor alpha (ERα) and thus influence estrogen-responsive gene expression [183]. Both proteins are expressed in the cytoplasm and in the nuclei of epithelial breast cancer MCF-7 cells and, together with 17beta-estradiol, they alter the level of hydrogen peroxide. However, estradiol was shown to rapidly decrease the expression of thioredoxin- interacting protein followed by an increase in the levels of cytosolic thioredoxin 1 (Txn 1) and mitochondrial thioredoxin 2 (Txn2) [53].

In turn, the changes in the intracellular redox state influence the function of estrogen receptor (ER) in estrogen-dependent cells. Addition of H_2O_2 at low concentrations down-regulated ER transcription activity, which was recovered by transfection of endogenous redox effector protein thioredoxin [83]. Furthermore, ROS induce activation of extracellular signal-regulated kinases 1/2 (ERK1/2) and protein kinase B (Akt), which were implicated in the phosphorylation of serine 118 and serine 167 on ERα, respectively [100, 242].Treatment of MCF-7 human breast cancer cells with glucose oxidase, which converts glucose present in the medium into d-glucono-1,5-lacton and hydrogen peroxide, induced transient phosphorylation of serine 118 and serine 167, which was enhanced in Her2 (ErbB2) over-expressing MCF-7 cells [233]. Inhibition of ERK1/2 decreased phosphorylation at both Ser 118 and Ser 167, while inhibition of Akt had no effect on phosphorylation at Ser167.

Ras and Redox state

Over-expression or activating mutation of Ras is described in almost 30% of all tumors [25, 227] Up to date, three members of the Ras family have been identified, namely H-Ras, K-Ras and N-Ras. Although these Ras isoforms function in similar ways, numerous evidences also support the distinct molecular functions of each Ras protein. Overexpression was originally reported in breast cancer [212, 211, 44]. Although mutations in Ras was considered a rare event (less than 5%) in development of breast cancer, activating mutation of K-Ras was observed at late stages of the development of the disease (tumors grade II and III) [212, 80, 152]. Activation of Ras leads to a constitutive activation of signal transduction molecules, including mitogen-activated protein kinase (MAPK), which become uncoupled from extracellular mitogenic stimuli resulting in an uncontrolled cellular proliferation and cell resistance to the oxidative stress. Increased expression of enzymes involved in regulation of intracellular redox environment catalase correlated well with elevated tolerance of K-Ras mutants to the hydrogen peroxide cytotoxicity and enhanced intracellular glutathione (GSH). Upregulation of a key enzyme of the pentose phosphate pathway in K-Ras mutants resulted in an increased production of redox equivalent NADPH required for anabolic processes as well as the reducing of oxidized GSH providing the growth advantages for cancer cells [185]. Transfection of epithelial breast cancer MCF-7 cells or normal breast epithelial cells (MCF-10A) with activated Ras resulted in increased tumorigenicity of MCF-7 cells and transformation of MCF-10A cells [7, 17]. Among the principal effectors of Ras are Raf and p110 (a catalytic subunit of PI3K), which are known to facilitate cellular transformation through increased cell survival and uncontrolled proliferation. The effects of Ras were linked to Rac-dependent regulation of superoxide production [93, 94]. A connection between Ras and Nox was later established when K-Ras(V12) was shown to induce Nox1, while suppression of NOX1 blocked Ras-dependent tumor formation [[151, 2]. Although Ras was implicated in generation of ROS, oxidants in turn affect Ras either via direct modification of H-Ras Cys118 or by influencing the GTP/GDP exchange through increased Ras GTP loading [116, 117, 118]. C118S Ras mutant was shown to block Ras and ERK1/2 activation and prevented phosphorylation of p38 MAPK and Akt, suggesting that oxidative modification of Ras is crucial for execution of Ras oncogenic activity [45, 240]. It was proposed that the loss of Prx1 in cancer cells resulting in more oxidized redox environment promotes oncogenic H-Ras or ErbB2/neuT-induced transforamation [31].

c-jun and c-fos

The sensitivity of c-fos and c-jun to redox state is due the presence of conserved cysteine residues in their DNA binding domains that allows to bind to DNA under reduced redox conditions. Shifting the redox potential to more oxidized state prevent c-fos/c-jun DNA biding due formation of disulphide bridge between cysteine residues within dimmer [14]. Upregulation of thiols or/and overexpression of thioredoxin activate c-fos and c-jun genes and consequently lead to the formation of c-jun/c-fos AP-1 dimers and therefore activation of AP-1 [148, 200]. It was reported that thioredoxin alone or in conjunction with the full thioredoxin system that comprises thioredoxin, thioredoxin reductase (TR), and NADPH upregulate AP1, probably through a direct interaction between Trx and the C-terminal of Jun activation domain-binding protein 1 (Jab1) [65, 22, 88]. DNA-binding activities of AP-1 families (including c-fos) are redox-regulated via Cys residues modification [19, 243]. Exposure to micromolar concentrations of hydrogen peroxide arrests cell cycle progression by blocking the expression of cyclin D1 via stabilization of c-fos, which prevents fra-1 from accessing chromatin [29, 182]. However, low concentrations of H2O2 (100 nM to 1 mM) enhanced c-fos expression, which was associated with an increase in cell proliferation [231]. Both increased ROS generation and suppressed ROS generation in the presence of antioxidants cause transient phosphorylation of Elk-1 kinase on identical sites in the Elk-1 transactivation domain leading to induction of c-fos [158]. Ras-dependent activation of ERK2 observed in response to redox state disturbances caused by hydrogen peroxide and antioxidants led to induction of Elk1 driven c-fos [79, 158]. Similar phenomena was observed under hypoxia which causes cellular redox imbalance. Therefore, changes in the cellular redox state to either more oxidized or reduced state regulate activation and deactivation of AP-1.

Metallothioneins and redox regulation of oncogenes

Metallothioneins (MTs) in couple with the glutathione and thioredoxin redox pairs protect cells against reactive and damaging effects of ROS, control the intracellular redox state pushing the redox environment toward a reduced state and protect easily oxidized molecules against oxidizing species including ROS. Numerous studies showed frequent downregulation of several isoforms of MT in hepatocellular carcinomas [90, 89, 52]. The suggested mechanism underlying the repression of MT was via hypermethylation of the MT promoter and PI3K/Akt induced dephosphorylation of the transcription factor CCAAT/enhancer-binding protein(C/EBP)-alpha [41, 52]. However, MT1 and abundantly expressed MT2A isoforms were shown to be highly expressed in human invasive ductal breast cancer reaching their highest levels in breast cancer cells with an invasive phenotype [164, 51, 99, 244]. Strong MT expression was observed in the majority of breast carcinoma cells and was strongly associated with the components of ductal carcinoma with low steroid receptor status [92]. Although the role of MT 1 and 2 in oncogenesis is not well understood, upregulation of MT in metastatic breast cancer cells strongly suggests a role on the part of MT in breast cancer cells survival and formation of invasive phenotype. High expression of MT associates with poor tumor grade, chemotherapy resistance with low recurrence-survival, which indicates an important role of MT in predicting breast cancer recurrence [99, 244]. Highly proliferative cancer cells show intense MT immunoreactivity [57]. In dividing cells, MT localization depends on the phase of the cell cycle progression [150, 50, 41]. Thus, during the G0 and G1 phases, MTs are primarily localized in the cytoplasm, while during the S phase they accumulate in the nucleus with subsequent redistribution back to the cytoplasm at G2/M [40, 150, 126]. Similar to MTs, Prx 1 is upregulated in cells entering the S phase of the cell cycle suggesting a requirement for highly reduced redox environment for proper regulation of replication [208]. On the contrary, silencing of MT2A gene in MCF-7 cells induced cell cycle arrest in the G1-phase via the ATM/Chk2/cdc25A pathway [127]. It was demonstrated that MT localizes in the nuclei in response to generation of ROS suggesting that MT enters the nuclei when it senses intracellular oxidation [216]. These data suggest that MT may have a fundamental impact on the cell cycle regulation.

MT was linked to expression of Bcl-2 and oncogene c-myc, both of which control cell cycle progression, apoptosis and angiogenesis [9, 225]. It was shown that suppression of MT induced growth inhibition, which was associated with decreased bcl-2 and c-myc and induced expression of c-fos and p53, whereas overexpression of cytoplasmic MT increased cell multiplication and reversed oncogene expression [1, 171, 218]. These data suggest that the pro-survival effects of MT 1 and 2 are mediated via Bcl-2 and c-myc regulation and that the interrelationships between MT and c-myc and bcl-2 implicate MT in the control of cell proliferation and thereby in the multistep process of oncogenesis. MT 1 and 2 were implicated in activation of cyclin D and stimulation of cycle progression [153, 42]. Furthermore, MT 1 and 2 stimulate expression of TGF-β, TGF-β-receptor, VEGF required for building the new vascular network that supplies both oxygen and nutrients to tumor cells [165].

In view of metallothionein ability to bind up to seven divalent metal ions such as zinc, MTs can regulate the DNA binding activity of tumor suppressor p53 through zinc transfer reaction [146]. Coordination of zinc by Cys176, His179, Cys238, and Cys24 residues is required for folding p53 into functional 'wild-type' conformation. However, the metal-free form of MT (apo-MT), which is usually present in the cell when it is transiently generated during protein synthesis, may disrupt the DNA binding activity of p53 by competing for zinc and/or through direct physical interaction between p53 and apo-metallothionein [146]. The complex might be formed by interaction between sulfhydryl groups of apo-MT and zinc ion of p53 since there is no interaction between MT and p53 [170]. Thus, apo-MT interaction with p53 may represent a significant regulatory mechanism by which metallothionein can impair tumor suppressor function thereby promoting tumorigenesis.

NADPH oxidases mediated regulation of oncogene signaling

Accumulation of oxidizing species such as reactive oxygen species shifts the cellular redox environment to a more oxidized state. The ROS mediated redox regulation may be direct or indirect, through redox modifications of proteins mainly at cysteine (Cys) residues. For instance, H_2O_2 mediated oxidation of cysteines from reduced-SH to oxidized -S-S- form and $O2^{\cdot -}$ affects the redox state of metal cofactors in kinases and phosphatases. In some cases, other amino acids can be oxidized, such as Tyr, Trp, and His. Redox modulations of kinases and phosphatases could influence the activities of cell cycle regulatory pathways, which in turn could regulate progression from one cell cycle phase to the next one [61]. ROS are continuously generated in living cells as by-products of aerobic respiration and various metabolic processes or are directly produced from oxygen. In mammalian cells, a major non-mitochondrial source of reactive oxygen species is NADPH oxidase (Nox) complexes. Nox family members are broadly distributed in many tissues, and are increasingly implicated in signal transduction processes and intercommunication. The Nox family consists of Nox from 1 to 5 and dual oxidases 1 and 2. Nox1 and 3 are similar to the phagocytic Nox2 and require formation of p22phox and p47phox (noxo1) complexes for their activation, whereas dual oxidases (DUOX) are targeted by dual oxidase maturation factors (DuoxA1 or DuoxA2). Each homologue of Nox enzymes exhibits a distinct cellular and tissue distribution pattern and shows distinct, specific functional roles in cells [115]. In terms of relationships with cancer development, elevated ROS production has been recognized in various malignant cells and Nox 1, 2, 4 and 5 have been implicated in survival of tumor cells and cell spreading process [27, 26, 223]. It was suggested that oncogenic Ras-MAPK signalling, in addition to Fra-1 activation, induces constitutive expression of Nox1, and then Nox1-generated ROS mediate Ras oncogene transformation characterized by amplified cell growth, altered cell morphology, anchorage-independent growth and tumorigenesis [151, 205, 94]. Nox1-induced ROS production is a necessary step for the signalling cascade that controls alteration of stress fibers and focal adhesions associated with Ras transformation [205]. This process involves oxidative inactivation of LMW-PTP and subsequent activation of p190RhoGAP, a direct target of the phosphatase. This causes down-regulation of Rho, leading to the dysregulation of action stress fibers and focal adhesions.

Overexpression of Nox1 increases the cell cycle progression by activatingAkt and maintaining AP-1 (fos-jun complex) dependent induction of cyclin D1 [182]. Cyclin D1 responds to changes in the cellular redox environment in each phase of cell cycle and its expression is regulated by the redox-sensitive transcriptional response of c-Myc and AP-1 binding to the promoter region of cyclin D1 [47, 12, 78, 161]. On the other hand, substantially enhanced generation of ROS resulting from expression of Nox4 or expression of Nox1 with Noxo1 and Noxa1 suppressed induction of cyclin D1 and cell proliferation [182]. Unlike other superoxide generating members of the Nox family, activated dual oxidase 1 (DUOXA1), which mainly produces hydrogen peroxide, diminished epithelial breast cancer cell proliferation through upregulation of p21$^{CIP1/WAF1}$ expression, which reflects substantial changes in expression of cyclin D1 [169].

Suppression of Nox1 activity does affect the responses of Akt or ERK. On the contrary, it has been shown that PX domains of p47phox and p40phox, the cytosolic components of Nox 1 to 4, recognize specific phosphoinositol lipids generated by PI3-kinase providing link between activation of PI3K/Akt signaling pathway and activation of Noxs and generation of ROS [102]. Therefore, persistent activation of PI3-kinase might itself be responsible for enhanced prolonged generation of ROS. However, inactivation of Akt by LY294002 also results in a significant increase in hydrogen peroxide generation associated with upregulation of DUOX maturation factors 1 (DUOXA1), which is required for functional reconstitution of the dual oxidase and accompanied by upregulation of cell cycle inhibitor p21$^{CIP1/WAF1}$ [180, 177].

Intracellular Redox regulation and cell cycle progression

Cell proliferation is crucial for maintenance of tissue homeostasis. The intracellular redox environment is of considerable significance for regulation of the cell cycle progression. The level of intracellular reduced glutathione (GSH), which reflects the cellular redox state, varied in each phase of the cell cycle [140]. In proliferating cells, glutathione is recruited into the nucleus at early phases of the cell cycle [136]. It was indicated that the GSH content is significantly higher in the G2/M phase compared to G1, suggesting a more reduced intracellular redox environment [47]. Experiments with thiol-antioxidant, N-acetyl-L-cysteine (NAC), showed that an increase in the levels of intracellular low-molecular-weight thiols, which indicates a shift in the intracellular redox state towards a more reduced environment, associates with a selective delay of the cell cycle progression and an increase in p27 and Rb hypophosphorylation [145, 143, 144]. Depletion of GSH by treating cells with buthionine-[S, R]-sulfoximine (BSO), a glutathione synthesis inhibitor, was shown to impair mitosis – promoting factor and induce cell cycle arrest at the G1/S transition phase accompanied by inactivation of Cdk1, Akt and ERK1/2 [186]. Redox perturbations initiated by diethylmaleate (DEM), a glutathione (GSH)-depleting agent, stimulated generation of ROS accompanied by increased expression of p21^{CIP1WAF1} [58]. Induction of p21$^{CIP1/WAF1}$ led to inhibition of kinase activities, accumulation of E2F/pRb complex and consequently cell cycle arrest. The Rb gene product acts as a transcriptional repressor of genes involved in the S phase progression by associating with E2F transcription factors [213, 149]. It was suggested that there is a possibility that regulation of redox-sensitive thiol-disulfide reactions at critical cysteine residues in cyclin D1, p27, and Rb proteins could act as "sulfhydryl switches" that modulate the cell cycle progression [145, 217]. Therefore, changes in the intracellular redox state provide a link between redox sensitive regulatory proteins in the G1 phase and transition from the G1 to S phase. Alternatively, ROS may either stimulate or inhibit cell division depending on their concentration and duration of exposure. Thus, sub–lethal doses of ROS (superoxide and hydrogen peroxide) added exogenously stimulate proliferation and initiate cell cycle progression in response to growth factors [214, 121].

In healthy epithelial cells, cell proliferation is tightly regulated. In cancer, regulation of cell proliferation is mismanaged in one way or another due to mutations or hyperactivation of genes that control the cell cycle machinery leading to autonomous cell cycle transition. The level of glutathione markedly increases in transformed cells characterized by uncontrolled cell proliferation [107]. It was shown that transformed cells were able to grow in the presence of GSH-depleting agent such as BSO, which inhibited activation of Mek1 and Akt, whereas control cells were not. However, enhanced generation of ROS associated with depletion of GSH was also implicated in activation of Cdk1, Akt and ERK and facilitation of cell cycle progression, that suggesting that the outcome of the effects of ROS depends on the overall intracellular redox potential. A critical regulator of the cellular redox state is Nrf2 transcription factor, which is required for constitutive and inducible expression of Gclm and Gclc involved in GSH biosynthesis. Furthermore, NRF2 regulates expressions of cellular detoxifying enzymes, such as NQO1, HO-1 and glutathione S-transferases [181]. Thus, Nrf2 deficiency led to alveolar epithelial cell G2/M cell cycle arrest accompanied by inactivation of Akt signaling and oxidative stress as a result of diminished level of GSH [187, 188]. Supplementation of Nrf2 deficient cells with GSH restored the cell cycle progression [188, 186], whereas supplementation with a thiol antioxidant and a precursor of intracellular glutathione, N-acetylcysteine, failed to restore the cell cycle progression but abolished the activation of ATM and p53.

Redox Regulation of Cyclin D1

Facilitation of the cell cycle progression and uncontrolled expression of cyclin D1 are common features in various types of cancer including breast cancer. The endogenous production of ROS operates as a key signaling process in the cascade of events leading to cell proliferation after stimulation with oncogenic Ras [94]. The cell cycle progression is a tightly regulated sequence of transitions from the G1 to the S phase and then to the G2 and M phases which involves sequential activation of cyclin-dependent kinase [159, 75, 82]. Progression from the G1 to the S phase is largely regulated by D-type cyclins, in particular cyclin D1, since inhibition of cyclin D1 expression prevents the transition from the G1 to the S phase, while its ectopic expression accelerate the transition to the S phase [203, 160, 190].

Up-regulation of cyclin D1 was reported in over 50% of human mammary carcinomas and was shown to be essential for development of breast cancer induced by c-neu and v-Ha-ras, but not by c-myc or Wnt-1 [28, 16, 54, 229, 247]. The lack of a dependence between c-myc and cyclin D1 in breast carcinomas most likely is due to the fact that c-myc regulate the cell cycle progression through sequestration of $p27^{KIP1}$ into cyclin D2-cdk complex [179, 23]. Ras stimulates both synthesis and assembly of cyclin D1 through the Raf1–MEK–ERK pathway and stabilizes cyclin D1 through PI3K–Akt–GSK-3 signaling pathway [6, 122, 74, 105, 4, 39].

A NAC-mediated increase in intracellular low-molecular-weight thiols in mouse embryonic fibroblast cell associates with suppression of cyclin D1, an increase in p27, and consequently a delay in the cell cycle progression from the G0/G1 to the S phase[145]. Similarly, in C6 glioma cells, treatment with NAC decreased proliferation by inducing the cell cycle arrest at the G0/G1 phase, up-regulating $p21^{CIP1/WAF1}$ expression, and lowering the activity of Akt and ERK [137]. A rapid and glutathione-independent decrease in intracellular oxidants was shown in cells treated with NAC. Menon and coauthors hypothesize that the intracellular redox state differentially affects the cell-cycle progression in nonmalignant versus malignant cells [143]. The authors showed that N-acetyl-L-cysteine, which increased the levels of intracellular small-molecular-weight thiols, glutathione and cysteine, induced G1 delays in nonmalignant human breast epithelial MCF-10A cells without causing any significant changes in the cell cycle progression in breast cancer epithelial MCF-7 and MDA-MB-231 cells. NAC-induced shifts toward a more reduced intracellular redox state led to a decrease in cyclin D1 and an increase in p27 protein in nonmalignant MCF-10A cells but not in MCF-7 and MDA-MB-231 cells.

Intriguingly, a thiol antioxidant, N-acetyl-L-cysteine, has been shown to inhibit progression from G1 to S via increased generation of superoxide anion-radicals [144]. It was shown that under physiological conditions, thiols may undergo one electron oxidation to generate thiyl radicals ($RS^{\cdot}$), which, in turn, can react with the thiolate anion (RS^{-}) to form the disulfide radical anion ($RSSR^{\cdot}$). In the presence of molecular oxygen, disulfide radical-anion yields disulfides (RSSR) and $O2^{\cdot}$ radical-anion [236]. Although NAC increased intracellular GSH and cysteine, changes in the intracellular thiol pools did not affect the NAC mediated cell cycle arrest since inhibition of GSH synthesis using buthionine-(S, R)-sulfoximine did not reverse the arrest [Menon, 2003 144 /id}. Furthermore, an increase in the cellular $O_2^{\cdot-}$ level associated with a decrease in cyclin D1 and activation of catalase and MnSOD but not CuZnSOD [144]. Co-treatment with catalase or overexpression of catalase did not avert the effect of NAC, whereas scavengers of superoxide reversed the NAC mediated suppression of cyclin D1. Therefore, these data suggest that regulation of cyclin D1 depends on concentration and type of reactive oxygen species and the overall intracellular redox potential.

Redox regulation of PI3K/PTEN/Akt signaling

Akt, a serine/threonine kinase, plays a distinguished role in regulation of cell proliferation and apoptosis. Akt is activated by sequential phosphorylation at Ser(473) by mTORC2 and at Thr(308) by PDK1. It was demonstrated that H_2O_2 stimulates an entrance of cytosolic Akt1 phosphorylated at Ser(473) to mitochondria, where it is further phosphorylated at Thr(308) by constitutive PDK1 leading to nuclear relocation and activation of genomic post-translational mechanisms regulating the cell proliferation [8]. However, high levels of H_2O_2 disrupt the Akt1-PDK1 interaction, which prevents nuclear accumulation of activated Akt and suppresses the cell proliferation. Retention of Akt1 Thr308 in mitochondria in the presence of high concentrations of H_2O_2 increases cytochrome c release to cytosol leading to apoptosis. Differential effects of high and low dosage of H_2O_2 on Akt1-PDK interaction were shown to be regulated by selective oxidation of Cys(310) to sulfenic or cysteic acids [8].

The tumor suppressor PTEN is a PtdIns(3,4,5)P(3) phosphatase that regulates cell migration, growth, and survival through direct antagonism of PI 3-kinase signaling by removing the 3'-phosphate of phosphoinositides. Phosphorylation of Akt by superoxide and hydrogen peroxide requires accumulation of 3'-phosphorylated phosphoinositides (PIP3) (PtdIns(3,4,5)P(3)) and inactivation of phosphatase PTEN [123, 125, 128]. PTEN oxidation in response to H_2O_2 enhances PI3K signaling leading to activation of Akt and facilitation of cell proliferation [46]. PTEN was shown to be the most sensitive to inactivation and regulation by oxidation among the members of the Protein Tyrosine Phosphatase (PTP) family of enzymes, which contain a nucleophilic reactive catalytic cysteine within a conserved motif that permits redox regulation of these enzymes [192]. PTEN is easily inactivated by hydrogen peroxide through reversible oxidation characterized by disulfide bond formation between the essential Cys(124) residue in the active site of PTEN and Cys(71) [123, 46, 192, 55]. The reduction of oxidized PTEN is predominantly mediated by thioredoxin; although PTEN oxidation can be also reversed by coexpression of H_2O_2-detoxifying enzyme catalase. In contrast to intramolecular disulfide formation in the presence of H_2O_2, S-nitrosothiols, S-nitrosocysteine (CSNO) and S-nitrosoglutathione (GSNO) reversibly oxidize PTEN through formation of a mixed disulfide [246]. Interestingly, $O_2^{\cdot-}$-mediated oxidation and consequent inhibition of the PTEN activity is due to S-nitrosylation of the protein [128]. PTEN is also sensitive to a lipid peroxidation by-product, 4-hydroxynonenal [196]. Increased accumulation of NADH caused by inhibition of mitochondrial respiration and hypoxia also inactivates PTEN through a redox modification mechanism, leading to Akt activation [174].

Redox adaptation to oncogenic mediated hypoxia and glycolysis

Coupled Glycolysis and oncogenesis

Clinical studies reveal a strong statistical association between low aerobic capacity and death from all causes. However, in cancer, reversion to anaerobic respiration or glycolysis for ATP production (the Warburg effect) underlies increased cancer cell survival. However, ATP generation through glycolysis is less efficient than through mitochondrial respiration. And yet, cancer cells with this metabolic disadvantage survive the competition with other cells and eventually develop resistance to radio- and chemotherapy. Respiration-deficient cells exhibit dependency on glycolysis, increased NADH, and activation of Akt, leading to advantage in hypoxia [174]. It was shown that the levels of phosphocholine, phosphoethanolamine, and glycerol derivatives of these metabolites were very low in human mammary epithelial cells and significantly less than that in human breast cancer cells (MCF-7 and T47D), while high energy phosphates and the rates of glucose consumption did not differ between control and cancer cells [219]. Overexpression of membrane glucose transporters is a common feature of malignant tumors including breast cancer and is associated with invasiveness. Statistically significant correlation was shown between negativity of estrogen receptor ER-alpha and high expression of glucose transporter GLUT-1 [119]. Similarly, estrogen receptor negative epithelial breast cancer MDA-MB-231 showed higher expression of GLUT1 than ER positive MCF-7 cells [120]. Furthermore, strong link was observed between plasma insulin and estrogen receptor (ER) negative breast cancer [84, 68].

The metabolic and morphological characterization of human epithelial breast cancer MCF-7 cells and noncancerous 48R human mammary epithelial cells (48R HMECs) showed that cancer cells exhibited a higher activity of the pentose phosphate pathway and a higher glucose consumption rate, suggesting high energy requirement and at the same time high efficiency of cancer cells [141]. Furthermore, although adriamycin increased the pentose phosphate shunt activity in both adriamycin-sensitive control and adriamycin-resistant breast cancer MCF-7 cells, hydrogen peroxide markedly stimulated pentose phosphate pathway only in adriamycin-resistant cells [245]. Estradiol was shown to increase the carbon flow through the pentose phosphate pathway and increases glutamine consumption that limited NADH recycle, resulting in NADH buildup in the cytosol [62].

Using human breast cancer cell model for brain metastasis based on circulating tumor cells from a breast cancer in immunodeficient mice, it was shown that the brain-derived cells exhibited increased expression of enzymes involved in glycolysis, tricarboxylic acid cycle, and oxidative phosphorylation pathways, as well as enhanced activation of the pentose phosphate pathway and the glutathione system [36]. The pentose phosphate pathway serves to generate NADPH, which is considered a primary reducing molecule for glutathione system [87, 86, 199]. The upregulation of this pathway coupled with enhanced glutathione reductase and glutathione S-transferase activities minimizes the oxidative stress caused by reactive oxygen species and maintains high GSH levels. These changes suggest that metastatic cells derive energy from glucose oxidation, which affects the cellular redox balance and explains a decrease in ROS production, enhanced tumor cell survival and proliferation, and enhanced resistance to drugs. It is not surprising that breast cancer cells show much lower activities of the mitochondrial respiratory chain complexes, cellular oxygen consumption, and ATP synthesis rates than normal mammary epithelial cells [135].

The transformed phenotype was characterized by increased catabolism of glucose by the pentose phosphate pathway and increased synthesis of NADPH, therefore reflecting the need of the transformed cells for the reducing power of NADPH to protect the rapidly proliferating cell from reactive oxygen species through keeping high level of glutathione [63, 64, 189]. These observations were consistent with alterations observed in the H-ras transformed MCF10-AT cells with increased metastatic potential. The changes in metastatic MCF10-CA1a cells and H-ras transformed MCF10-AT cells included an increase in the level of succinate that reflects pseudohypoxic conditions and promotes a tumorigenic phenotype [189, 201, 178]. At the same time, during the exponential growth phase of K-ras transformed cells, oxidative phosphorylation was down-regulated due to a decrease in complex I activity [15]. The increase in the TCA cycle flux could increase the succinate level, which in turn could increase the level of HIF-1a and thus contribute to tumorigenicity [132].

Lu and coauthors declared a two-step metabolic progression hypothesis during mammary tumor progression [133]. According to Lu and coauthors, the first step towards acquisition of tumorigenicity includes enhanced glycolysis, pentose phosphate pathway, and fatty acid synthesis, as well as decreased GSH/GSSG redox pool; the second step is correlated with the gain of the general metastatic ability and includes further changes in glycolysis and tricarboxylic acid cycle (TCA cycle), and further depletion of glutathione species. The decrease in reduced and oxidized forms of glutathione in tumorigenic lines described by Lu and coauthors imply that GSH depletion may facilitate the transformation by predisposing cells to ROS mediated mutations. However, their data showed continued GSH/GSSG depletion in cells gaining metastatic phenotype thus contradicting the evidence showing that metastatic tumor cells were very well equipped to cope with redox stresses, which allows cancer cells to resist drug treatment. Upregulation of the pentose phosphate pathway was responsible for accumulation of NADPH and therefore increased level of reduced form of GSH. However, there is a possibility of the third way of adaptation that stimulates oxidative glutaminolysis via dysregulation of PI3K/Akt/mTOR network and MYC, and, along with anoxic glutaminolysis, provides pyruvate, lactate, and high quantities of NADPH to meet the metabolic requirements associated with a high growth rate [209]. This may explain the high survival rate and proliferation potential of metastatic cancer cells.

Antidiabetic drug metformin, which induces glucose homeostasis through promotion of glucose uptake, significantly reduces the risk of breast cancer. The effects of metformin include suppression of Her2 (ErbB2) protein level via direct (AMPK-independent) inactivation of p70S6K1 and inhibition of cyclin D1 resulting in breast cancer cell cycle arrest and growth inhibition [224, 18]. Co-incubation with N-acetylcysteine dramatically enhanced the ability of metformin to decrease HER2 expression indicating the importance of mTOR/p70S6K1-sensed ROS status for mediating the anti-oncogenic effects of metformin.

Changes in redox regulation in response to oxygen tension (hypoxia)

It is known that hypoxia induces glycolysis and a glucose consumption increase in metastatic breast cancer cells versus non-invasive cells [71, 72]. The master regulators of aerobic glycolysis that support the metabolic autonomy of tumor cells are members of the PI3K/Akt/mTOR pathway, hypoxia-inducible factor HIF-1α and c-myc. Akt activation increases the expression of glucose transporter and glycolytic enzymes. Interestingly, HIF-1α is stabilized by the activation of the PI3K/Akt/mTOR pathway [98, 138]. Activation of HIF-1α leads to increased expression of glucose transporters and glycolytic enzymes to meet the need of highly proliferating tumor cells [72]. In early stages of carcinogenesis, hypoxia triggers an increase in the glycolytic flux and induces changes that preserve the advantages of HIF-1α mediated signalling under conditions of low pO_2 [72]. c-Myc stimulates the expression of glycolytic enzymes [204] and glutaminolysis, therefore providing tumor cells with large quantities of NADPH needed. The uncoupling of glycolysis from oxidative metabolism is predictable when local pO_2 is low to avoid oxidative stress by reducing electron flux through oxidative phosphorylation [60]. Expressions of both pyruvate dehydrogenase kinase and lactate dehydrogenase A are induced in response to an increase in HIF-1a activity. The conversion of pyruvate into lactate ensuring NAD^+ regeneration is induced by a variety of oncogenes, including c-myc [204].

Hypoxia has been reported to lower the intracellular redox potential that shifts into more oxidized state and is followed by induction of AP-1 transcription factor DNA-binding activity [13]. PDTC, a potent reducing agent, activates the AP-1 transcription factor in HeLa cells, and increases accumulation of c-jun mRNA in these cells. However, hypoxic stimulation of AP-1 transcription factor binding to its cognate DNA sequence did not lead to activation of c-jun gene. Activation of PI3K/Akt pathway is crucial for HIF-1α stabilization during early hypoxia [156]. However, prolonged hypoxia inactivates Akt and activates GSK3β, which results in decreased HIF-1α accumulation. An increased expression of mitochondrial thioredoxin peroxidase and peroxiredoxin 3 (Prx3) reduces the amount of H_2O_2 and improves glucose homeostasis characterized by reduced levels of glucose and increased glucose clearance [37]. On the contrary, suppression of HIF-1α expression increases generation of reactive oxygen species and shifts the intracellular redox environment into more oxidizing state characterized by low GSH/GSSG ratio and low NADPH level [77]. Both thioredoxin and thioredoxin reductase are upregulated under hypoxic conditions. Thioredoxin-1 increases both aerobic and hypoxia-induced HIF-α, which controls angiogenesis [234]. Inhibitors of Trx-1 were shown to decrease HIF-1α protein levels and RNA and protein expression of the HIF-1α target genes in the MDA-MB-468 and MCF-7 breast cancer cell lines [101, 235, 157]. The inhibition of HIF-1α by Trx-1 inhibitors led to a decrease in HIF-1-trans-activating activity and VEGF formation. Suppression of thioredoxin reductase was also associated with increases in VEGF and VEGF receptor expression, cell migration, proliferation and angiogenesis [215].

Conclusion

Perturbations in the redox environment increase the damaging effects of oxidants, in particular, ROS produced in the course of cellular aerobic metabolism, which results in oxidative DNA damage, induction of mutation and stimulation of oncogenic redox sensitive pathways. Oxidative modification of thiol groups of Ras, c-fos and c-jun dysregulate genes that control cell proliferation and survival leading to uncontrolled epithelial mammary cell growth. Although continuously exposed to increased level of oxidants, highly aggressive metastatic breast cancer cells show adaptation to oxidative stress by maintaining a highly reducing intracellular redox state.

References

Abdel-Mageed A.; Agrawal K.C. (1997). Antisense down-regulation of metallothionein induces growth arrest and apoptosis in human breast carcinoma cells. Cancer Gene Ther., 4, 199-207.

Adachi Y.; Shibai Y.; Mitsushita J.; Shang W.H.; Hirose K.; Kamata T. (2008). Oncogenic Ras upregulates NADPH oxidase 1 gene expression through MEK-ERK-dependent phosphorylation of GATA-6. Oncogene, 27, 4921-4932.

Agnoletto M.H.; Guecheva T.N.; Donde F.; de Oliveira A.F.; Franke F.; Cassini C.; Salvador M.; Henriques J.A.; Saffi J. (2007). Association of low repair efficiency with high hormone receptors expression and SOD activity in breast cancer patients. Clin.Biochem., 40, 1252-1258.

Ahamed S.; Foster J.S.; Bukovsky A.; Wimalasena J. (2001). Signal transduction through the Ras/Erk pathway is essential for the mycoestrogen zearalenone-induced cell-cycle progression in MCF-7 cells. Mol.Carcinog., 30, 88-98.

Ahsan M.K.; Masutani H.; Yamaguchi Y.; Kim Y.C.; Nosaka K.; Matsuoka M.; Nishinaka Y.; Maeda M.; Yodoi J. (2006). Loss of interleukin-2-dependency in HTLV-I-infected T cells on gene silencing of thioredoxin-binding protein-2. Oncogene, 25, 2181-2191.

Albanese C.; Johnson J.; Watanabe G.; Eklund N.; Vu D.; Arnold A.; Pestell R.G. (1995). Transforming p21ras mutants and c-Ets-2 activate the cyclin D1 promoter through distinguishable regions. J.Biol.Chem., 270, 23589-23597.

Albini A.; Graf J.; Kitten G.T.; Kleinman H.K.; Martin G.R.; Veillette A.; Lippman M.E. (1986). 17 beta-estradiol regulates and v-Ha-ras transfection constitutively enhances MCF7 breast cancer cell interactions with basement membrane. Proc.Natl.Acad.Sci.U.S.A, 83, 8182-8186.

Antico A., V; Galli S.; Franco M.C.; Lam P.Y.; Cadenas E.; Carreras M.C.; Poderoso J.J. (2009). Akt1 intramitochondrial cycling is a crucial step in the redox modulation of cell cycle progression. PLoS.One., 4, e7523.

Arvanitis C.; Felsher D.W. (2006). Conditional transgenic models define how MYC initiates and maintains tumorigenesis. Semin.Cancer Biol., 16, 313-317.

Bader A.G.; Kang S.; Vogt P.K. (2006). Cancer-specific mutations in PIK3CA are oncogenic in vivo. Proc.Natl.Acad.Sci.U.S.A, 103, 1475-1479.

Baker A.; Payne C.M.; Briehl M.M.; Powis G. (1997). Thioredoxin, a gene found overexpressed in human cancer, inhibits apoptosis in vitro and in vivo. Cancer Res., 57, 5162-5167.

Bakiri L.; Lallemand D.; Bossy-Wetzel E.; Yaniv M. (2000). Cell cycle-dependent variations in c-Jun and JunB phosphorylation: a role in the control of cyclin D1 expression. EMBO J., 19, 2056-2068.

Bandyopadhyay R.S.; Phelan M.; Faller D.V. (1995). Hypoxia induces AP-1-regulated genes and AP-1 transcription factor binding in human endothelial and other cell types. Biochim.Biophys.Acta, 1264, 72-78.

Bannister A.J.; Cook A.; Kouzarides T. (1991). In vitro DNA binding activity of Fos/Jun and BZLF1 but not C/EBP is affected by redox changes. Oncogene, 6, 1243-1250.

Baracca A.; Chiaradonna F.; Sgarbi G.; Solaini G.; Alberghina L.; Lenaz G. (2010). Mitochondrial Complex I decrease is responsible for bioenergetic dysfunction in K-ras transformed cells. Biochim.Biophys.Acta, 1797, 314-323.

Bartkova J.; Lukas J.; Muller H.; Lutzhoft D.; Strauss M.; Bartek J. (1994). Cyclin D1 protein expression and function in human breast cancer. Int.J.Cancer, 57, 353-361.

Basolo F.; Elliott J.; Tait L.; Chen X.Q.; Maloney T.; Russo I.H.; Pauley R.; Momiki S.; Caamano J.; Klein-Szanto A.J.; . (1991). Transformation of human breast epithelial cells by c-Ha-ras oncogene. Mol.Carcinog., 4, 25-35.

Ben S., I; Laurent K.; Loubat A.; Giorgetti-Peraldi S.; Colosetti P.; Auberger P.; Tanti J.F.; Marchand-Brustel Y.; Bost F. (2008). The antidiabetic drug metformin exerts an antitumoral effect in vitro and in vivo through a decrease of cyclin D1 level. Oncogene, 27, 3576-3586.

Benezra R. (1994). An intermolecular disulfide bond stabilizes E2A homodimers and is required for DNA binding at physiological temperatures. Cell, 79, 1057-1067.

Berggren M.; Gallegos A.; Gasdaska J.R.; Gasdaska P.Y.; Warneke J.; Powis G. (1996). Thioredoxin and thioredoxin reductase gene expression in human tumors and cell lines, and the effects of serum stimulation and hypoxia. Anticancer Res., 16, 3459-3466.

Berggren M.I.; Husbeck B.; Samulitis B.; Baker A.F.; Gallegos A.; Powis G. (2001). Thioredoxin peroxidase-1 (peroxiredoxin-1) is increased in thioredoxin-1 transfected cells and results in enhanced protection against apoptosis caused by hydrogen peroxide but not by other agents including dexamethasone, etoposide, and doxorubicin. Arch.Biochem.Biophys., 392, 103-109.

Bloomfield K.L.; Osborne S.A.; Kennedy D.D.; Clarke F.M.; Tonissen K.F. (2003). Thioredoxin-mediated redox control of the transcription factor Sp1 and regulation of the thioredoxin gene promoter. Gene, 319, 107-116.

Bodrug S.E.; Warner B.J.; Bath M.L.; Lindeman G.J.; Harris A.W.; Adams J.M. (1994). Cyclin D1 transgene impedes lymphocyte maturation and collaborates in lymphomagenesis with the myc gene. EMBO J., 13, 2124-2130.

Bolton J.L.; Thatcher G.R. (2008). Potential mechanisms of estrogen quinone carcinogenesis. Chem.Res.Toxicol., 21, 93-101.

Bos J.L. (1988). The ras gene family and human carcinogenesis. Mutat.Res., 195, 255-271.

Brar S.S.; Corbin Z.; Kennedy T.P.; Hemendinger R.; Thornton L.; Bommarius B.; Arnold R.S.; Whorton A.R.; Sturrock A.B.; Huecksteadt T.P.; Quinn M.T.; Krenitsky K.; Ardie K.G.; Lambeth J.D.; Hoidal J.R. (2003). NOX5 NAD(P)H oxidase regulates growth and apoptosis in DU 145 prostate cancer cells. Am.J.Physiol Cell Physiol, 285, C353-C369.

Brar S.S.; Kennedy T.P.; Sturrock A.B.; Huecksteadt T.P.; Quinn M.T.; Whorton A.R.; Hoidal J.R. (2002). An NAD(P)H oxidase regulates growth and transcription in melanoma cells. Am.J.Physiol Cell Physiol, 282, C1212-C1224.

Buckley M.F.; Sweeney K.J.; Hamilton J.A.; Sini R.L.; Manning D.L.; Nicholson R.I.; deFazio A.; Watts C.K.; Musgrove E.A.; Sutherland R.L. (1993). Expression and amplification of cyclin genes in human breast cancer. Oncogene, 8, 2127-2133.

Burch P.M.; Yuan Z.; Loonen A.; Heintz N.H. (2004). An extracellular signal-regulated kinase 1- and 2-dependent program of chromatin trafficking of c-Fos and Fra-1 is required for cyclin D1 expression during cell cycle reentry. Mol.Cell Biol., 24, 4696-4709.

Butler L.M.; Zhou X.; Xu W.S.; Scher H.I.; Rifkind R.A.; Marks P.A.; Richon V.M. (2002). The histone deacetylase inhibitor SAHA arrests cancer cell growth, up-regulates thioredoxin-binding protein-2, and down-regulates thioredoxin. Proc.Natl.Acad.Sci.U.S.A, 99, 11700-11705.

Cao J.; Schulte J.; Knight A.; Leslie N.R.; Zagozdzon A.; Bronson R.; Manevich Y.; Beeson C.; Neumann C.A. (2009). Prdx1 inhibits tumorigenesis via regulating PTEN/AKT activity. EMBO J., 28, 1505-1517.

Carpten J.D.; Faber A.L.; Horn C.; Donoho G.P.; Briggs S.L.; Robbins C.M.; Hostetter G.; Boguslawski S.; Moses T.Y.; Savage S.; Uhlik M.; Lin A.; Du J.; Qian Y.W.; Zeckner D.J.; Tucker-Kellogg G.; Touchman J.; Patel K.; Mousses S.; Bittner M.; Schevitz R.; Lai M.H.; Blanchard K.L.; Thomas J.E. (2007). A transforming mutation in the pleckstrin homology domain of AKT1 in cancer. Nature, 448, 439-444.

Castrellon A.B.; Gluck S. (2008). Chemoprevention of breast cancer. Expert.Rev.Anticancer Ther., 8, 443-452.

Cha M.K.; Suh K.H.; Kim I.H. (2009). Overexpression of peroxiredoxin I and thioredoxin1 in human breast carcinoma. J.Exp.Clin.Cancer Res., 28, 93.

Chae H.Z.; Chung S.J.; Rhee S.G. (1994). Thioredoxin-dependent peroxide reductase from yeast. J.Biol.Chem., 269, 27670-27678.

Chen E.I.; Hewel J.; Krueger J.S.; Tiraby C.; Weber M.R.; Kralli A.; Becker K.; Yates J.R., III; Felding-Habermann B. (2007). Adaptation of energy metabolism in breast cancer brain metastases. Cancer Res., 67, 1472-1486.

Chen L.; Na R.; Gu M.; Salmon A.B.; Liu Y.; Liang H.; Qi W.; Van Remmen H.; Richardson A.; Ran Q. (2008). Reduction of mitochondrial H2O2 by overexpressing peroxiredoxin 3 improves glucose tolerance in mice. Aging Cell, 7, 866-878.

Chen X.P.; Liu S.; Tang W.X.; Chen Z.W. (2007). Nuclear thioredoxin-1 is required to suppress cisplatin-mediated apoptosis of MCF-7 cells. Biochem.Biophys.Res.Commun., 361, 362-366.

Cheng M.; Sexl V.; Sherr C.J.; Roussel M.F. (1998). Assembly of cyclin D-dependent kinase and titration of p27Kip1 regulated by mitogen-activated protein kinase kinase (MEK1). Proc.Natl.Acad.Sci.U.S.A, 95, 1091-1096.

Cherian M.G.; Apostolova M.D. (2000). Nuclear localization of metallothionein during cell proliferation and differentiation. Cell Mol.Biol.(Noisy.-le-grand), 46, 347-356.

Cherian M.G.; Jayasurya A.; Bay B.H. (2003). Metallothioneins in human tumors and potential roles in carcinogenesis. Mutat.Res., 533, 201-209.

Cherian M.G.; Kang Y.J. (2006). Metallothionein and liver cell regeneration. Exp.Biol.Med.(Maywood.), 231, 138-144.

Chua P.J.; Lee E.H.; Yu Y.; Yip G.W.; Tan P.H.; Bay B.H. (2010). Silencing the Peroxiredoxin III gene inhibits cell proliferation in breast cancer. Int.J.Oncol., 36, 359-364.

Clair T.; Miller W.R.; Cho-Chung Y.S. (1987). Prognostic significance of the expression of a ras protein with a molecular weight of 21,000 by human breast cancer. Cancer Res., 47, 5290-5293.

Clavreul N.; Bachschmid M.M.; Hou X.; Shi C.; Idrizovic A.; Ido Y.; Pimentel D.; Cohen R.A. (2006). S-glutathiolation of p21ras by peroxynitrite mediates endothelial insulin resistance caused by oxidized low-density lipoprotein. Arterioscler.Thromb.Vasc.Biol., 26, 2454-2461.

Connor K.M.; Subbaram S.; Regan K.J.; Nelson K.K.; Mazurkiewicz J.E.; Bartholomew P.J.; Aplin A.E.; Tai Y.T.; Aguirre-Ghiso J.; Flores S.C.; Melendez J.A. (2005). Mitochondrial H2O2 regulates the angiogenic phenotype via PTEN oxidation. J.Biol.Chem., 280, 16916-16924.

Conour J.E.; Graham W.V.; Gaskins H.R. (2004). A combined in vitro/bioinformatic investigation of redox regulatory mechanisms governing cell cycle progression. Physiol Genomics, 18, 196-205.

Cotgreave I.A.; Gerdes R.G. (1998). Recent trends in glutathione biochemistry--glutathione-protein interactions: a molecular link between oxidative stress and cell proliferation? Biochem.Biophys.Res.Commun., 242, 1-9.

Coughlin S.S.; Ekwueme D.U. (2009). Breast cancer as a global health concern. Cancer Epidemiol., 33, 315-318.

Coyle P.; Philcox J.C.; Carey L.C.; Rofe A.M. (2002). Metallothionein: the multipurpose protein. Cell Mol.Life Sci., 59, 627-647.

Dalgin G.S.; Drever M.; Williams T.; King T.; DeLisi C.; Liou L.S. (2008). Identification of novel epigenetic markers for clear cell renal cell carcinoma. J.Urol., 180, 1126-1130.

Datta J.; Majumder S.; Kutay H.; Motiwala T.; Frankel W.; Costa R.; Cha H.C.; MacDougald O.A.; Jacob S.T.; Ghoshal K. (2007). Metallothionein expression is suppressed in primary human hepatocellular carcinomas and is mediated through inactivation of CCAAT/enhancer binding protein alpha by phosphatidylinositol 3-kinase signaling cascade. Cancer Res., 67, 2736-2746.

Deroo B.J.; Hewitt S.C.; Peddada S.D.; Korach K.S. (2004). Estradiol regulates the thioredoxin antioxidant system in the mouse uterus. Endocrinology, 145, 5485-5492.

Dickson C.; Fantl V.; Gillett C.; Brookes S.; Bartek J.; Smith R.; Fisher C.; Barnes D.; Peters G. (1995). Amplification of chromosome band 11q13 and a role for cyclin D1 in human breast cancer. Cancer Lett., 90, 43-50.

Downes C.P.; Walker S.; McConnachie G.; Lindsay Y.; Batty I.H.; Leslie N.R. (2004). Acute regulation of the tumour suppressor phosphatase, PTEN, by anionic lipids and reactive oxygen species. Biochem.Soc.Trans., 32, 338-342.

Droge W. (2002). Free radicals in the physiological control of cell function. Physiol Rev., 82, 47-95.

Dutsch-Wicherek M.; Sikora J.; Tomaszewska R. (2008). The possible biological role of metallothionein in apoptosis. Front Biosci., 13, 4029-4038.

Esposito F.; Russo L.; Chirico G.; Ammendola R.; Russo T.; Cimino F. (2001). Regulation of p21waf1/cip1 expression by intracellular redox conditions. IUBMB.Life, 52, 67-70.

Feigin M.E.; Muthuswamy S.K. (2009). ErbB receptors and cell polarity: new pathways and paradigms for understanding cell migration and invasion. Exp.Cell Res., 315, 707-716.

Feron O. (2009). Pyruvate into lactate and back: from the Warburg effect to symbiotic energy fuel exchange in cancer cells. Radiother.Oncol., 92, 329-333.

Finkel T. (2001). Reactive oxygen species and signal transduction. IUBMB.Life, 52, 3-6.

Forbes N.S.; Meadows A.L.; Clark D.S.; Blanch H.W. (2006). Estradiol stimulates the biosynthetic pathways of breast cancer cells: detection by metabolic flux analysis. Metab Eng, 8, 639-652.

Frederiks W.M.; Bosch K.S.; De Jong J.S.; Van Noorden C.J. (2003). Post-translational regulation of glucose-6-phosphate dehydrogenase activity in (pre)neoplastic lesions in rat liver. J.Histochem.Cytochem., 51, 105-112.

Frederiks W.M.; Vizan P.; Bosch K.S.; Vreeling-Sindelarova H.; Boren J.; Cascante M. (2008). Elevated activity of the oxidative and non-oxidative pentose phosphate pathway in (pre)neoplastic lesions in rat liver. Int.J.Exp.Pathol., 89, 232-240.

Freemerman A.J.; Gallegos A.; Powis G. (1999). Nuclear factor kappaB transactivation is increased but is not involved in the proliferative effects of thioredoxin overexpression in MCF-7 breast cancer cells. Cancer Res., 59, 4090-4094.

Freemerman A.J.; Powis G. (2000). A redox-inactive thioredoxin reduces growth and enhances apoptosis in WEHI7.2 cells. Biochem.Biophys.Res.Commun., 274, 136-141.

Gallegos A.; Gasdaska J.R.; Taylor C.W.; Paine-Murrieta G.D.; Goodman D.; Gasdaska P.Y.; Berggren M.; Briehl M.M.; Powis G. (1996). Transfection with human thioredoxin increases cell proliferation and a dominant-negative mutant thioredoxin reverses the transformed phenotype of human breast cancer cells. Cancer Res., 56, 5765-5770.

Gamayunova V.B.; Bobrov Y.; Tsyrlina E.V.; Evtushenko T.P.; Berstein L.M. (1997). Comparative study of blood insulin levels in breast and endometrial cancer patients. Neoplasma, 44, 123-126.

Gasdaska J.R.; Berggren M.; Powis G. (1995). Cell growth stimulation by the redox protein thioredoxin occurs by a novel helper mechanism. Cell Growth Differ., 6, 1643-1650.

Gasdaska P.Y.; Oblong J.E.; Cotgreave I.A.; Powis G. (1994). The predicted amino acid sequence of human thioredoxin is identical to that of the autocrine growth factor human adult T-cell derived factor (ADF): thioredoxin mRNA is elevated in some human tumors. Biochim.Biophys.Acta, 1218, 292-296.

Gatenby R.A.; Gillies R.J. (2004). Why do cancers have high aerobic glycolysis? Nat.Rev.Cancer, 4, 891-899.

Gillies R.J.; Robey I.; Gatenby R.A. (2008). Causes and consequences of increased glucose metabolism of cancers. J.Nucl.Med., 49 Suppl 2, 24S-42S.

Gordon L.I.; Burke M.A.; Singh A.T.; Prachand S.; Lieberman E.D.; Sun L.; Naik T.J.; Prasad S.V.; Ardehali H. (2009). Blockade of the erbB2 receptor induces cardiomyocyte death through mitochondrial and reactive oxygen species-dependent pathways. J.Biol.Chem., 284, 2080-2087.

Gotoh J.; Obata M.; Yoshie M.; Kasai S.; Ogawa K. (2003). Cyclin D1 over-expression correlates with beta-catenin activation, but not with H-ras mutations, and phosphorylation of Akt, GSK3 beta and ERK1/2 in mouse hepatic carcinogenesis. Carcinogenesis, 24, 435-442.

Grana X.; Reddy E.P. (1995). Cell cycle control in mammalian cells: role of cyclins, cyclin dependent kinases (CDKs), growth suppressor genes and cyclin-dependent kinase inhibitors (CKIs). Oncogene, 11, 211-219.

Green R.M.; Graham M.; O'Donovan M.R.; Chipman J.K.; Hodges N.J. (2006). Subcellular compartmentalization of glutathione: correlations with parameters of oxidative stress related to genotoxicity. Mutagenesis, 21, 383-390.

Guo S.; Miyake M.; Liu K.J.; Shi H. (2009). Specific inhibition of hypoxia inducible factor 1 exaggerates cell injury induced by in vitro ischemia through deteriorating cellular redox environment. J.Neurochem., 108, 1309-1321.

Guttridge D.C.; Albanese C.; Reuther J.Y.; Pestell R.G.; Baldwin A.S., Jr. (1999). NF-kappaB controls cell growth and differentiation through transcriptional regulation of cyclin D1. Mol.Cell Biol., 19, 5785-5799.

Guyton K.Z.; Liu Y.; Gorospe M.; Xu Q.; Holbrook N.J. (1996). Activation of mitogen-activated protein kinase by H2O2. Role in cell survival following oxidant injury. J.Biol.Chem., 271, 4138-4142.

Hachisuga T.; Tsujioka H.; Horiuchi S.; Udou T.; Emoto M.; Kawarabayashi T. (2005). K-ras mutation in the endometrium of tamoxifen-treated breast cancer patients, with a comparison of tamoxifen and toremifene. Br.J.Cancer, 92, 1098-1103.

Halliwell B., Gutteridge J.M.C. (2007) 'Free Radicals in Biology and Medicine.' (Clarendon Press, Oxford, UK.:

Hartwell L.H.; Weinert T.A. (1989). Checkpoints: controls that ensure the order of cell cycle events. Science, 246, 629-634.

Hayashi S.; Hajiro-Nakanishi K.; Makino Y.; Eguchi H.; Yodoi J.; Tanaka H. (1997). Functional modulation of estrogen receptor by redox state with reference to thioredoxin as a mediator. Nucleic Acids Res., 25, 4035-4040.

Hirose K.; Toyama T.; Iwata H.; Takezaki T.; Hamajima N.; Tajima K. (2003). Insulin, insulin-like growth factor-I and breast cancer risk in Japanese women. Asian Pac.J.Cancer Prev., 4, 239-246.

Hirotsu S.; Abe Y.; Okada K.; Nagahara N.; Hori H.; Nishino T.; Hakoshima T. (1999). Crystal structure of a multifunctional 2-Cys peroxiredoxin heme-binding protein 23 kDa/proliferation-associated gene product. Proc.Natl.Acad.Sci.U.S.A, 96, 12333-12338.

Holmgren A. (1989). Thioredoxin and glutaredoxin systems. J.Biol.Chem., 264, 13963-13966.

HORECKER B.L. (1962). Alternative pathways of carbohydrate metabolism in relation to evolutionary development. Comp Biochem.Physiol, 4, 363-369.

Hwang C.Y.; Ryu Y.S.; Chung M.S.; Kim K.D.; Park S.S.; Chae S.K.; Chae H.Z.; Kwon K.S. (2004). Thioredoxin modulates activator protein 1 (AP-1) activity and p27Kip1 degradation through direct interaction with Jab1. Oncogene, 23, 8868-8875.

Iizuka N.; Oka M.; Yamada-Okabe H.; Mori N.; Tamesa T.; Okada T.; Takemoto N.; Sakamoto K.; Hamada K.; Ishitsuka H.; Miyamoto T.; Uchimura S.; Hamamoto Y. (2005). Self-organizing-map-based molecular signature representing the development of hepatocellular carcinoma. FEBS Lett., 579, 1089-1100.

Iizuka N.; Oka M.; Yamada-Okabe H.; Mori N.; Tamesa T.; Okada T.; Takemoto N.; Tangoku A.; Hamada K.; Nakayama H.; Miyamoto T.; Uchimura S.; Hamamoto Y. (2002). Comparison of gene expression profiles between hepatitis B virus- and hepatitis C virus-infected hepatocellular carcinoma by oligonucleotide microarray data on the basis of a supervised learning method. Cancer Res., 62, 3939-3944.

Ikenoue T.; Kanai F.; Hikiba Y.; Obata T.; Tanaka Y.; Imamura J.; Ohta M.; Jazag A.; Guleng B.; Tateishi K.; Asaoka Y.; Matsumura M.; Kawabe T.; Omata M. (2005). Functional analysis of PIK3CA gene mutations in human colorectal cancer. Cancer Res., 65, 4562-4567.

Ioachim E.; Kamina S.; Demou A.; Kontostolis M.; Lolis D.; Agnantis N.J. (1999). Immunohistochemical localization of metallothionein in human breast cancer in comparison with cathepsin D, stromelysin-1, CD44, extracellular matrix components, P53, Rb, C-erbB-2, EGFR, steroid receptor content and proliferation. Anticancer Res., 19, 2133-2139.

Irani K.; Goldschmidt-Clermont P.J. (1998). Ras, superoxide and signal transduction. Biochem.Pharmacol., 55, 1339-1346.

Irani K.; Xia Y.; Zweier J.L.; Sollott S.J.; Der C.J.; Fearon E.R.; Sundaresan M.; Finkel T.; Goldschmidt-Clermont P.J. (1997). Mitogenic signaling mediated by oxidants in Ras-transformed fibroblasts. Science, 275, 1649-1652.

Isakoff S.J.; Engelman J.A.; Irie H.Y.; Luo J.; Brachmann S.M.; Pearline R.V.; Cantley L.C.; Brugge J.S. (2005). Breast cancer-associated PIK3CA mutations are oncogenic in mammary epithelial cells. Cancer Res., 65, 10992-11000.

Jang H.H.; Lee K.O.; Chi Y.H.; Jung B.G.; Park S.K.; Park J.H.; Lee J.R.; Lee S.S.; Moon J.C.; Yun J.W.; Choi Y.O.; Kim W.Y.; Kang J.S.; Cheong G.W.; Yun D.J.; Rhee S.G.; Cho M.J.; Lee S.Y. (2004). Two enzymes in one; two yeast peroxiredoxins display oxidative stress-dependent switching from a peroxidase to a molecular chaperone function. Cell, 117, 625-635.

Jemal A.; Murray T.; Samuels A.; Ghafoor A.; Ward E.; Thun M.J. (2003). Cancer statistics, 2003. CA Cancer J.Clin., 53, 5-26.

Jiang B.H.; Jiang G.; Zheng J.Z.; Lu Z.; Hunter T.; Vogt P.K. (2001). Phosphatidylinositol 3-kinase signaling controls levels of hypoxia-inducible factor 1. Cell Growth Differ., 12, 363-369.

Jin R.; Chow V.T.; Tan P.H.; Dheen S.T.; Duan W.; Bay B.H. (2002). Metallothionein 2A expression is associated with cell proliferation in breast cancer. Carcinogenesis, 23, 81-86.

Joel P.B.; Smith J.; Sturgill T.W.; Fisher T.L.; Blenis J.; Lannigan D.A. (1998). pp90rsk1 regulates estrogen receptor-mediated transcription through phosphorylation of Ser-167. Mol.Cell Biol., 18, 1978-1984.

Jones D.T.; Pugh C.W.; Wigfield S.; Stevens M.F.; Harris A.L. (2006). Novel thioredoxin inhibitors paradoxically increase hypoxia-inducible factor-alpha expression but decrease functional transcriptional activity, DNA binding, and degradation. Clin.Cancer Res., 12, 5384-5394.

Kanai F.; Liu H.; Field S.J.; Akbary H.; Matsuo T.; Brown G.E.; Cantley L.C.; Yaffe M.B. (2001). The PX domains of p47phox and p40phox bind to lipid products of PI(3)K. Nat.Cell Biol., 3, 675-678.

Karihtala P.; Mantyniemi A.; Kang S.W.; Kinnula V.L.; Soini Y. (2003). Peroxiredoxins in breast carcinoma. Clin.Cancer Res., 9, 3418-3424.

Kawahara N.; Tanaka T.; Yokomizo A.; Nanri H.; Ono M.; Wada M.; Kohno K.; Takenaka K.; Sugimachi K.; Kuwano M. (1996). Enhanced coexpression of thioredoxin and high mobility group protein 1 genes in human hepatocellular carcinoma and the possible association with decreased sensitivity to cisplatin. Cancer Res., 56, 5330-5333.

Kerkhoff E.; Rapp U.R. (1997). Induction of cell proliferation in quiescent NIH 3T3 cells by oncogenic c-Raf-1. Mol.Cell Biol., 17, 2576-2586.

Khan S.; Kumagai T.; Vora J.; Bose N.; Sehgal I.; Koeffler P.H.; Bose S. (2004). PTEN promoter is methylated in a proportion of invasive breast cancers. Int.J.Cancer, 112, 407-410.

Kim S.J.; Kim H.G.; Lim H.W.; Park E.H.; Lim C.J. (2005). Up-regulation of glutathione biosynthesis in NIH3T3 cells transformed with the ETV6-NTRK3 gene fusion. Mol.Cells, 19, 131-136.

Kim S.J.; Miyoshi Y.; Taguchi T.; Tamaki Y.; Nakamura H.; Yodoi J.; Kato K.; Noguchi S. (2005). High thioredoxin expression is associated with resistance to docetaxel in primary breast cancer. Clin.Cancer Res., 11, 8425-8430.

Kim S.K.; Yang J.W.; Kim M.R.; Roh S.H.; Kim H.G.; Lee K.Y.; Jeong H.G.; Kang K.W. (2008). Increased expression of Nrf2/ARE-dependent anti-oxidant proteins in tamoxifen-resistant breast cancer cells. Free Radic.Biol.Med., 45, 537-546.

Koc A.; Mathews C.K.; Wheeler L.J.; Gross M.K.; Merrill G.F. (2006). Thioredoxin is required for deoxyribonucleotide pool maintenance during S phase. J.Biol.Chem., 281, 15058-15063.

Kuljaca S.; Liu T.; Dwarte T.; Kavallaris M.; Haber M.; Norris M.D.; Martin-Caballero J.; Marshall G.M. (2009). The cyclin-dependent kinase inhibitor, p21(WAF1), promotes angiogenesis by repressing gene transcription of thioredoxin-binding protein 2 in cancer cells. Carcinogenesis, 30, 1865-1871.

Kurebayashi J. (2001). Biological and clinical significance of HER2 overexpression in breast cancer. Breast Cancer, 8, 45-51.

Kwon J.; Lee S.R.; Yang K.S.; Ahn Y.; Kim Y.J.; Stadtman E.R.; Rhee S.G. (2004). Reversible oxidation and inactivation of the tumor suppressor PTEN in cells stimulated with peptide growth factors. Proc.Natl.Acad.Sci.U.S.A, 101, 16419-16424.

Lam J.B.; Chow K.H.; Xu A.; Lam K.S.; Liu J.; Wong N.S.; Moon R.T.; Shepherd P.R.; Cooper G.J.; Wang Y. (2009). Adiponectin haploinsufficiency promotes mammary tumor development in MMTV-PyVT mice by modulation of phosphatase and tensin homolog activities. PLoS.One., 4, e4968.

Lambeth J.D. (2004). NOX enzymes and the biology of reactive oxygen. Nat.Rev.Immunol., 4, 181-189.

Lander H.M.; Hajjar D.P.; Hempstead B.L.; Mirza U.A.; Chait B.T.; Campbell S.; Quilliam L.A. (1997). A molecular redox switch on p21(ras). Structural basis for the nitric oxide-p21(ras) interaction. J.Biol.Chem., 272, 4323-4326.

Lander H.M.; Ogiste J.S.; Pearce S.F.; Levi R.; Novogrodsky A. (1995). Nitric oxide-stimulated guanine nucleotide exchange on p21ras. J.Biol.Chem., 270, 7017-7020.

Lander H.M.; Ogiste J.S.; Teng K.K.; Novogrodsky A. (1995). p21ras as a common signaling target of reactive free radicals and cellular redox stress. J.Biol.Chem., 270, 21195-21198.

Laudanski P.; Koda M.; Kozlowski L.; Swiatecka J.; Wojtukiewicz M.; Sulkowski S.; Wolczynski S. (2004). Expression of glucose transporter GLUT-1 and estrogen receptors ER-alpha and ER-beta in human breast cancer. Neoplasma, 51, 164-168.

Laudanski P.; Swiatecka J.; Kovalchuk O.; Wolczynski S. (2003). Expression of GLUT1 gene in breast cancer cell lines MCF-7 and MDA-MB-231. Ginekol.Pol., 74, 782-785.

Laurent A.; Nicco C.; Chereau C.; Goulvestre C.; Alexandre J.; Alves A.; Levy E.; Goldwasser F.; Panis Y.; Soubrane O.; Weill B.; Batteux F. (2005). Controlling tumor growth by modulating endogenous production of reactive oxygen species. Cancer Res., 65, 948-956.

Lavoie J.N.; Rivard N.; L'Allemain G.; Pouyssegur J. (1996). A temporal and biochemical link between growth factor-activated MAP kinases, cyclin D1 induction and cell cycle entry. Prog.Cell Cycle Res., 2, 49-58.

Lee S.R.; Yang K.S.; Kwon J.; Lee C.; Jeong W.; Rhee S.G. (2002). Reversible inactivation of the tumor suppressor PTEN by H2O2. J.Biol.Chem., 277, 20336-20342.

Lei K.; Townsend D.M.; Tew K.D. (2008). Protein cysteine sulfinic acid reductase (sulfiredoxin) as a regulator of cell proliferation and drug response. Oncogene, 27, 4877-4887.

Leslie N.R.; Bennett D.; Lindsay Y.E.; Stewart H.; Gray A.; Downes C.P. (2003). Redox regulation of PI 3-kinase signalling via inactivation of PTEN. EMBO J., 22, 5501-5510.

Levadoux-Martin M.; Hesketh J.E.; Beattie J.H.; Wallace H.M. (2001). Influence of metallothionein-1 localization on its function. Biochem.J., 355, 473-479.

Lim D.; Jocelyn K.M.; Yip G.W.; Bay B.H. (2009). Silencing the Metallothionein-2A gene inhibits cell cycle progression from G1- to S-phase involving ATM and cdc25A signaling in breast cancer cells. Cancer Lett., 276, 109-117.

Lim S.; Clement M.V. (2007). Phosphorylation of the survival kinase Akt by superoxide is dependent on an ascorbate-reversible oxidation of PTEN. Free Radic.Biol.Med., 42, 1178-1192.

Lincoln D.T.; Al Yatama F.; Mohammed F.M.; Al Banaw A.G.; Al Bader M.; Burge M.; Sinowatz F.; Singal P.K. (2010). Thioredoxin and thioredoxin reductase expression in thyroid cancer depends on tumour aggressiveness. Anticancer Res., 30, 767-775.

Lincoln D.T.; Ali Emadi E.M.; Tonissen K.F.; Clarke F.M. (2003). The thioredoxin-thioredoxin reductase system: over-expression in human cancer. Anticancer Res., 23, 2425-2433.

Liu H.; Nishitoh H.; Ichijo H.; Kyriakis J.M. (2000). Activation of apoptosis signal-regulating kinase 1 (ASK1) by tumor necrosis factor receptor-associated factor 2 requires prior dissociation of the ASK1 inhibitor thioredoxin. Mol.Cell Biol., 20, 2198-2208.

Lu H.; Forbes R.A.; Verma A. (2002). Hypoxia-inducible factor 1 activation by aerobic glycolysis implicates the Warburg effect in carcinogenesis. J.Biol.Chem., 277, 23111-23115.

Lu X.; Bennet B.; Mu E.; Rabinowitz J.; Kang Y. (2010). Metabolomic changes accompanying transformation and acquisition of metastatic potential in a syngeneic mouse mammary tumor model. J.Biol.Chem., 285, 9317-9321.

Ma X.; Karra S.; Lindner D.J.; Hu J.; Reddy S.P.; Kimchi A.; Yodoi J.; Kalvakolanu D.V. (2001). Thioredoxin participates in a cell death pathway induced by interferon and retinoid combination. Oncogene, 20, 3703-3715.

Ma Y.; Bai R.K.; Trieu R.; Wong L.J. (2010). Mitochondrial dysfunction in human breast cancer cells and their transmitochondrial cybrids. Biochim.Biophys.Acta, 1797, 29-37.

Markovic J.; Borras C.; Ortega A.; Sastre J.; Vina J.; Pallardo F.V. (2007). Glutathione is recruited into the nucleus in early phases of cell proliferation. J.Biol.Chem., 282, 20416-20424.

Martin V.; Herrera F.; Garcia-Santos G.; Antolin I.; Rodriguez-Blanco J.; Rodriguez C. (2007). Signaling pathways involved in antioxidant control of glioma cell proliferation. Free Radic.Biol.Med., 42, 1715-1722.

Martinive P.; Defresne F.; Quaghebeur E.; Daneau G.; Crokart N.; Gregoire V.; Gallez B.; Dessy C.; Feron O. (2009). Impact of cyclic hypoxia on HIF-1alpha regulation in endothelial cells--new insights for anti-tumor treatments. FEBS J., 276, 509-518.

Mates J.M.; Sanchez-Jimenez F.M. (2000). Role of reactive oxygen species in apoptosis: implications for cancer therapy. Int.J.Biochem.Cell Biol., 32, 157-170.

Mauro F.; Grasso A.; Tolmach L.J. (1969). Variations in sulfhydryl, disulfide, and protein content during synchronous and asynchronous growth of HeLa cells. Biophys.J., 9, 1377-1397.

Meadows A.L.; Kong B.; Berdichevsky M.; Roy S.; Rosiva R.; Blanch H.W.; Clark D.S. (2008). Metabolic and morphological differences between rapidly proliferating cancerous and normal breast epithelial cells. Biotechnol.Prog., 24, 334-341.

Meister A.; Anderson M.E. (1983). Glutathione. Annu.Rev.Biochem., 52, 711-760.

Menon S.G.; Coleman M.C.; Walsh S.A.; Spitz D.R.; Goswami P.C. (2005). Differential susceptibility of nonmalignant human breast epithelial cells and breast cancer cells to thiol antioxidant-induced G(1)-delay. Antioxid.Redox.Signal., 7, 711-718.

Menon S.G.; Sarsour E.H.; Kalen A.L.; Venkataraman S.; Hitchler M.J.; Domann F.E.; Oberley L.W.; Goswami P.C. (2007). Superoxide signaling mediates N-acetyl-L-cysteine-induced G1 arrest: regulatory role of cyclin D1 and manganese superoxide dismutase. Cancer Res., 67, 6392-6399.

Menon S.G.; Sarsour E.H.; Spitz D.R.; Higashikubo R.; Sturm M.; Zhang H.; Goswami P.C. (2003). Redox regulation of the G1 to S phase transition in the mouse embryo fibroblast cell cycle. Cancer Res., 63, 2109-2117.

Meplan C.; Richard M.J.; Hainaut P. (2000). Metalloregulation of the tumor suppressor protein p53: zinc mediates the renaturation of p53 after exposure to metal chelators in vitro and in intact cells. Oncogene, 19, 5227-5236.

Meredith M.J.; Cusick C.L.; Soltaninassab S.; Sekhar K.S.; Lu S.; Freeman M.L. (1998). Expression of Bcl-2 increases intracellular glutathione by inhibiting methionine-dependent GSH efflux. Biochem.Biophys.Res.Commun., 248, 458-463.

Meyer M.; Schreck R.; Baeuerle P.A. (1993). H2O2 and antioxidants have opposite effects on activation of NF-kappa B and AP-1 in intact cells: AP-1 as secondary antioxidant-responsive factor. EMBO J., 12, 2005-2015.

Michalides R.J.; van de B.M.; Balm F. (2002). Defects in G1-S cell cycle control in head and neck cancer: a review. Head Neck, 24, 694-704.

Miles A.T.; Hawksworth G.M.; Beattie J.H.; Rodilla V. (2000). Induction, regulation, degradation, and biological significance of mammalian metallothioneins. Crit Rev.Biochem.Mol.Biol., 35, 35-70.

Mitsushita J.; Lambeth J.D.; Kamata T. (2004). The superoxide-generating oxidase Nox1 is functionally required for Ras oncogene transformation. Cancer Res., 64, 3580-3585.

Miyakis S.; Sourvinos G.; Spandidos D.A. (1998). Differential expression and mutation of the ras family genes in human breast cancer. Biochem.Biophys.Res.Commun., 251, 609-612.

Miyashita H.; Sato Y. (2005). Metallothionein 1 is a downstream target of vascular endothelial zinc finger 1 (VEZF1) in endothelial cells and participates in the regulation of angiogenesis. Endothelium, 12, 163-170.

Mobley J.A.; Brueggemeier R.W. (2004). Estrogen receptor-mediated regulation of oxidative stress and DNA damage in breast cancer. Carcinogenesis, 25, 3-9.

Mochizuki M.; Kwon Y.W.; Yodoi J.; Masutani H. (2009). Thioredoxin regulates cell cycle via the ERK1/2-cyclin D1 pathway. Antioxid.Redox.Signal., 11, 2957-2971.

Mottet D.; Dumont V.; Deccache Y.; Demazy C.; Ninane N.; Raes M.; Michiels C. (2003). Regulation of hypoxia-inducible factor-1alpha protein level during hypoxic conditions by the phosphatidylinositol 3-kinase/Akt/glycogen synthase kinase 3beta pathway in HepG2 cells. J.Biol.Chem., 278, 31277-31285.

Mukherjee A.; Westwell A.D.; Bradshaw T.D.; Stevens M.F.; Carmichael J.; Martin S.G. (2005). Cytotoxic and antiangiogenic activity of AW464 (NSC 706704), a novel thioredoxin inhibitor: an in vitro study. Br.J.Cancer, 92, 350-358.

Muller J.M.; Cahill M.A.; Rupec R.A.; Baeuerle P.A.; Nordheim A. (1997). Antioxidants as well as oxidants activate c-fos via Ras-dependent activation of extracellular-signal-regulated kinase 2 and Elk-1. Eur.J.Biochem., 244, 45-52.

Muller R.; Mumberg D.; Lucibello F.C. (1993). Signals and genes in the control of cell-cycle progression. Biochim.Biophys.Acta, 1155, 151-179.

Musgrove E.A.; Lee C.S.; Buckley M.F.; Sutherland R.L. (1994). Cyclin D1 induction in breast cancer cells shortens G1 and is sufficient for cells arrested in G1 to complete the cell cycle. Proc.Natl.Acad.Sci.U.S.A, 91, 8022-8026.

Nagata D.; Suzuki E.; Nishimatsu H.; Satonaka H.; Goto A.; Omata M.; Hirata Y. (2001). Transcriptional activation of the cyclin D1 gene is mediated by multiple cis-elements, including SP1 sites and a cAMP-responsive element in vascular endothelial cells. J.Biol.Chem., 276, 662-669.

Nakamura H.; Bai J.; Nishinaka Y.; Ueda S.; Sasada T.; Ohshio G.; Imamura M.; Takabayashi A.; Yamaoka Y.; Yodoi J. (2000). Expression of thioredoxin and glutaredoxin, redox-regulating proteins, in pancreatic cancer. Cancer Detect.Prev., 24, 53-60.

Nano J.L.; Czerucka D.; Menguy F.; Rampal P. (1989). Effect of selenium on the growth of three human colon cancer cell lines. Biol.Trace Elem.Res., 20, 31-43.

Nguyen A.; Jing Z.; Mahoney P.S.; Davis R.; Sikka S.C.; Agrawal K.C.; Abdel-Mageed A.B. (2000). In vivo gene expression profile analysis of metallothionein in renal cell carcinoma. Cancer Lett., 160, 133-140.

Nielsen A.E.; Bohr A.; Penkowa M. (2007). The Balance between Life and Death of Cells: Roles of Metallothioneins. Biomark.Insights., 1, 99-111.

Nishinaka Y.; Masutani H.; Nakamura H.; Yodoi J. (2001). Regulatory roles of thioredoxin in oxidative stress-induced cellular responses. Redox.Rep., 6, 289-295.

Nishinaka Y.; Masutani H.; Oka S.; Matsuo Y.; Yamaguchi Y.; Nishio K.; Ishii Y.; Yodoi J. (2004). Importin alpha1 (Rch1) mediates nuclear translocation of thioredoxin-binding protein-2/vitamin D(3)-up-regulated protein 1. J.Biol.Chem., 279, 37559-37565.

Noh D.Y.; Ahn S.J.; Lee R.A.; Kim S.W.; Park I.A.; Chae H.Z. (2001). Overexpression of peroxiredoxin in human breast cancer. Anticancer Res., 21, 2085-2090.

Ostrakhovitch E.A.; Li S.S. (2010). NIP1/DUOXA1 expression in epithelial breast cancer cells: regulation of cell adhesion and actin dynamics. Breast Cancer Res.Treat., 119, 773-786.

Ostrakhovitch E.A.; Olsson P.E.; Jiang S.; Cherian M.G. (2006). Interaction of metallothionein with tumor suppressor p53 protein. FEBS Lett., 580, 1235-1238.

Ostrakhovitch E.A.; Olsson P.E.; von Hofsten J.; Cherian M.G. (2007). P53 mediated regulation of metallothionein transcription in breast cancer cells. J.Cell Biochem., 102, 1571-1583.

Parsonage D.; Youngblood D.S.; Sarma G.N.; Wood Z.A.; Karplus P.A.; Poole L.B. (2005). Analysis of the link between enzymatic activity and oligomeric state in AhpC, a bacterial peroxiredoxin. Biochemistry, 44, 10583-10592.

Parsons D.W.; Jones S.; Zhang X.; Lin J.C.; Leary R.J.; Angenendt P.; Mankoo P.; Carter H.; Siu I.M.; Gallia G.L.; Olivi A.; McLendon R.; Rasheed B.A.; Keir S.; Nikolskaya T.; Nikolsky Y.; Busam D.A.; Tekleab H.; Diaz L.A., Jr.; Hartigan J.; Smith D.R.; Strausberg R.L.; Marie S.K.; Shinjo S.M.; Yan H.; Riggins G.J.; Bigner D.D.; Karchin R.; Papadopoulos N.; Parmigiani G.; Vogelstein B.; Velculescu V.E.; Kinzler K.W. (2008). An integrated genomic analysis of human glioblastoma multiforme. Science, 321, 1807-1812.

Pelicano H.; Xu R.H.; Du M.; Feng L.; Sasaki R.; Carew J.S.; Hu Y.; Ramdas L.; Hu L.; Keating M.J.; Zhang W.; Plunkett W.; Huang P. (2006). Mitochondrial respiration defects in cancer cells cause activation of Akt survival pathway through a redox-mediated mechanism. J.Cell Biol., 175, 913-923.

Pfeifer G.P.; Besaratinia A. (2009). Mutational spectra of human cancer. Hum.Genet., 125, 493-506.

Phalen T.J.; Weirather K.; Deming P.B.; Anathy V.; Howe A.K.; van d., V; Jonsson T.J.; Poole L.B.; Heintz N.H. (2006). Oxidation state governs structural transitions in peroxiredoxin II that correlate with cell cycle arrest and recovery. J.Cell Biol., 175, 779-789.

Poh T.W.; Pervaiz S. (2005). LY294002 and LY303511 sensitize tumor cells to drug-induced apoptosis via intracellular hydrogen peroxide production independent of the phosphoinositide 3-kinase-Akt pathway. Cancer Res., 65, 6264-6274.

Pollard P.J.; Briere J.J.; Alam N.A.; Barwell J.; Barclay E.; Wortham N.C.; Hunt T.; Mitchell M.; Olpin S.; Moat S.J.; Hargreaves I.P.; Heales S.J.; Chung Y.L.; Griffiths J.R.; Dalgleish A.; McGrath J.A.; Gleeson M.J.; Hodgson S.V.; Poulsom R.; Rustin P.; Tomlinson I.P. (2005). Accumulation of Krebs cycle intermediates and over-expression of HIF1alpha in tumours which result from germline FH and SDH mutations. Hum.Mol.Genet., 14, 2231-2239.

Prall O.W.; Rogan E.M.; Musgrove E.A.; Watts C.K.; Sutherland R.L. (1998). c-Myc or cyclin D1 mimics estrogen effects on cyclin E-Cdk2 activation and cell cycle reentry. Mol.Cell Biol., 18, 4499-4508.

Ramanathan B.; Jan K.Y.; Chen C.H.; Hour T.C.; Yu H.J.; Pu Y.S. (2005). Resistance to paclitaxel is proportional to cellular total antioxidant capacity. Cancer Res., 65, 8455-8460.

Ramos-Gomez M.; Kwak M.K.; Dolan P.M.; Itoh K.; Yamamoto M.; Talalay P.; Kensler T.W. (2001). Sensitivity to carcinogenesis is increased and chemoprotective efficacy of enzyme inducers is lost in nrf2 transcription factor-deficient mice. Proc.Natl.Acad.Sci.U.S.A, 98, 3410-3415.

Ranjan P.; Anathy V.; Burch P.M.; Weirather K.; Lambeth J.D.; Heintz N.H. (2006). Redox-dependent expression of cyclin D1 and cell proliferation by Nox1 in mouse lung epithelial cells. Antioxid.Redox.Signal., 8, 1447-1459.

Rao A.K.; Ziegler Y.S.; McLeod I.X.; Yates J.R.; Nardulli A.M. (2009). Thioredoxin and thioredoxin reductase influence estrogen receptor alpha-mediated gene expression in human breast cancer cells. J.Mol.Endocrinol., 43, 251-261.

Ravi D.; Muniyappa H.; Das K.C. (2005). Endogenous thioredoxin is required for redox cycling of anthracyclines and p53-dependent apoptosis in cancer cells. J.Biol.Chem., 280, 40084-40096.

Recktenwald C.V.; Kellner R.; Lichtenfels R.; Seliger B. (2008). Altered detoxification status and increased resistance to oxidative stress by K-ras transformation. Cancer Res., 68, 10086-10093.

Reddy N.M.; Kleeberger S.R.; Bream J.H.; Fallon P.G.; Kensler T.W.; Yamamoto M.; Reddy S.P. (2008). Genetic disruption of the Nrf2 compromises cell-cycle progression by impairing GSH-induced redox signaling. Oncogene, 27, 5821-5832.

Reddy N.M.; Kleeberger S.R.; Cho H.Y.; Yamamoto M.; Kensler T.W.; Biswal S.; Reddy S.P. (2007). Deficiency in Nrf2-GSH signaling impairs type II cell growth and enhances sensitivity to oxidants. Am.J.Respir.Cell Mol.Biol., 37, 3-8.

Reddy N.M.; Kleeberger S.R.; Yamamoto M.; Kensler T.W.; Scollick C.; Biswal S.; Reddy S.P. (2007). Genetic dissection of the Nrf2-dependent redox signaling-regulated transcriptional programs of cell proliferation and cytoprotection. Physiol Genomics, 32, 74-81.

Richardson A.D.; Yang C.; Osterman A.; Smith J.W. (2008). Central carbon metabolism in the progression of mammary carcinoma. Breast Cancer Res.Treat., 110, 297-307.

Roberts J.M.; Koff A.; Polyak K.; Firpo E.; Collins S.; Ohtsubo M.; Massague J. (1994). Cyclins, Cdks, and cyclin kinase inhibitors. Cold Spring Harb.Symp.Quant.Biol., 59, 31-38.

Romieu-Mourez R.; Landesman-Bollag E.; Seldin D.C.; Sonenshein G.E. (2002). Protein kinase CK2 promotes aberrant activation of nuclear factor-kappaB, transformed phenotype, and survival of breast cancer cells. Cancer Res., 62, 6770-6778.

Ross S.H.; Lindsay Y.; Safrany S.T.; Lorenzo O.; Villa F.; Toth R.; Clague M.J.; Downes C.P.; Leslie N.R. (2007). Differential redox regulation within the PTP superfamily. Cell Signal., 19, 1521-1530.

Rudin C.M.; Yang Z.; Schumaker L.M.; VanderWeele D.J.; Newkirk K.; Egorin M.J.; Zuhowski E.G.; Cullen K.J. (2003). Inhibition of glutathione synthesis reverses Bcl-2-mediated cisplatin resistance. Cancer Res., 63, 312-318.

Rzymowska J.; Dyrda Z. (1993). Enzyme activities and level of SH groups in breast carcinomas. Int.J.Biochem., 25, 1331-1334.

Sadeq V.; Isar N.; Manoochehr T. (2010). Association of sporadic breast cancer with PTEN/MMAC1/TEP1 promoter hypermethylation. Med.Oncol.

Salsman S.J.; Hensley K.; Floyd R.A. (2005). Sensitivity of protein tyrosine phosphatase activity to the redox environment, cytochrome C, and microperoxidase. Antioxid.Redox.Signal., 7, 1078-1088.

Samuels Y.; Ericson K. (2006). Oncogenic PI3K and its role in cancer. Curr.Opin.Oncol., 18, 77-82.

Samuels Y.; Wang Z.; Bardelli A.; Silliman N.; Ptak J.; Szabo S.; Yan H.; Gazdar A.; Powell S.M.; Riggins G.J.; Willson J.K.; Markowitz S.; Kinzler K.W.; Vogelstein B.; Velculescu V.E. (2004). High frequency of mutations of the PIK3CA gene in human cancers. Science, 304, 554.

Schafer F.Q.; Buettner G.R. (2001). Redox environment of the cell as viewed through the redox state of the glutathione disulfide/glutathione couple. Free Radic.Biol.Med., 30, 1191-1212.

Schenk H.; Klein M.; Erdbrugger W.; Droge W.; Schulze-Osthoff K. (1994). Distinct effects of thioredoxin and antioxidants on the activation of transcription factors NF-kappa B and AP-1. Proc.Natl.Acad.Sci.U.S.A, 91, 1672-1676.

Selak M.A.; Armour S.M.; MacKenzie E.D.; Boulahbel H.; Watson D.G.; Mansfield K.D.; Pan Y.; Simon M.C.; Thompson C.B.; Gottlieb E. (2005). Succinate links TCA cycle dysfunction to oncogenesis by inhibiting HIF-alpha prolyl hydroxylase. Cancer Cell, 7, 77-85.

Sharma S.V.; Bell D.W.; Settleman J.; Haber D.A. (2007). Epidermal growth factor receptor mutations in lung cancer. Nat.Rev.Cancer, 7, 169-181.

Sherr C.J. (1993). Mammalian G1 cyclins. Cell, 73, 1059-1065.

Shim H.; Dolde C.; Lewis B.C.; Wu C.S.; Dang G.; Jungmann R.A.; Dalla-Favera R.; Dang C.V. (1997). c-Myc transactivation of LDH-A: implications for tumor metabolism and growth. Proc.Natl.Acad.Sci.U.S.A, 94, 6658-6663.

Shinohara M.; Shang W.H.; Kubodera M.; Harada S.; Mitsushita J.; Kato M.; Miyazaki H.; Sumimoto H.; Kamata T. (2007). Nox1 redox signaling mediates oncogenic Ras-induced disruption of stress fibers and focal adhesions by down-regulating Rho. J.Biol.Chem., 282, 17640-17648.

Siegel P.M.; Ryan E.D.; Cardiff R.D.; Muller W.J. (1999). Elevated expression of activated forms of Neu/ErbB-2 and ErbB-3 are involved in the induction of mammary tumors in transgenic mice: implications for human breast cancer. EMBO J., 18, 2149-2164.

Sjoblom T.; Jones S.; Wood L.D.; Parsons D.W.; Lin J.; Barber T.D.; Mandelker D.; Leary R.J.; Ptak J.; Silliman N.; Szabo S.; Buckhaults P.; Farrell C.; Meeh P.; Markowitz S.D.; Willis J.; Dawson D.; Willson J.K.; Gazdar A.F.; Hartigan J.; Wu L.; Liu C.; Parmigiani G.; Park B.H.; Bachman K.E.; Papadopoulos N.; Vogelstein B.; Kinzler K.W.; Velculescu V.E. (2006). The consensus coding sequences of human breast and colorectal cancers. Science, 314, 268-274.

Smith-Pearson P.S.; Kooshki M.; Spitz D.R.; Poole L.B.; Zhao W.; Robbins M.E. (2008). Decreasing peroxiredoxin II expression decreases glutathione, alters cell cycle distribution, and sensitizes glioma cells to ionizing radiation and H(2)O(2). Free Radic.Biol.Med., 45, 1178-1189.

Smolkova K.; Plecita-Hlavata L.; Bellance N.; Benard G.; Rossignol R.; Jezek P. (2010). Waves of gene regulation suppress and then restore oxidative phosphorylation in cancer cells. Int.J.Biochem.Cell Biol.

Somerville J.E.; Clarke L.A.; Biggart J.D. (1992). c-erbB-2 overexpression and histological type of in situ and invasive breast carcinoma. J.Clin.Pathol., 45, 16-20.

Spandidos D.A. (1987). Oncogene activation in malignant transformation: a study of H-ras in human breast cancer. Anticancer Res., 7, 991-996.

Spandidos D.A.; Pintzas A.; Kakkanas A.; Yiagnisis M.; Mahera H.; Patra E.; Agnantis N.J. (1987). Elevated expression of the myc gene in human benign and malignant breast lesions compared to normal tissue. Anticancer Res., 7, 1299-1304.

Stevaux O.; Dyson N.J. (2002). A revised picture of the E2F transcriptional network and RB function. Curr.Opin.Cell Biol., 14, 684-691.

Storz P. (2005). Reactive oxygen species in tumor progression. Front Biosci., 10, 1881-1896.

Streicher K.L.; Sylte M.J.; Johnson S.E.; Sordillo L.M. (2004). Thioredoxin reductase regulates angiogenesis by increasing endothelial cell-derived vascular endothelial growth factor. Nutr.Cancer, 50, 221-231.

Takahashi Y.; Ogra Y.; Suzuki K.T. (2005). Nuclear trafficking of metallothionein requires oxidation of a cytosolic partner. J.Cell Physiol, 202, 563-569.

Taya Y. (1997). RB kinases and RB-binding proteins: new points of view. Trends Biochem.Sci., 22, 14-17.

Theocharis S.E.; Margeli A.P.; Koutselinis A. (2003). Metallothionein: a multifunctional protein from toxicity to cancer. Int.J.Biol.Markers, 18, 162-169.

Ting Y.L.; Sherr D.; Degani H. (1996). Variations in energy and phospholipid metabolism in normal and cancer human mammary epithelial cells. Anticancer Res., 16, 1381-1388.

Turunen N.; Karihtala P.; Mantyniemi A.; Sormunen R.; Holmgren A.; Kinnula V.L.; Soini Y. (2004). Thioredoxin is associated with proliferation, p53 expression and negative estrogen and progesterone receptor status in breast carcinoma. APMIS, 112, 123-132.

Ueno M.; Matsutani Y.; Nakamura H.; Masutani H.; Yagi M.; Yamashiro H.; Kato H.; Inamoto T.; Yamauchi A.; Takahashi R.; Yamaoka Y.; Yodoi J. (2000). Possible association of thioredoxin and p53 in breast cancer. Immunol.Lett., 75, 15-20.

Ullmann G.M.; Knapp E.W. (1999). Electrostatic models for computing protonation and redox equilibria in proteins. Eur.Biophys.J., 28, 533-551.

Vaquero E.C.; Edderkaoui M.; Pandol S.J.; Gukovsky I.; Gukovskaya A.S. (2004). Reactive oxygen species produced by NAD(P)H oxidase inhibit apoptosis in pancreatic cancer cells. J.Biol.Chem., 279, 34643-34654.

Vazquez-Martin A.; Oliveras-Ferraros C.; Menendez J.A. (2009). The antidiabetic drug metformin suppresses HER2 (erbB-2) oncoprotein overexpression via inhibition of the mTOR effector p70S6K1 in human breast carcinoma cells. Cell Cycle, 8, 88-96.

Vita M.; Henriksson M. (2006). The Myc oncoprotein as a therapeutic target for human cancer. Semin.Cancer Biol., 16, 318-330.

Voehringer D.W.; Meyn R.E. (1998). Reversing drug resistance in bcl-2-expressing tumor cells by depleting glutathione. Drug Resist.Updat., 1, 345-351.

von Lintig F.C.; Dreilinger A.D.; Varki N.M.; Wallace A.M.; Casteel D.E.; Boss G.R. (2000). Ras activation in human breast cancer. Breast Cancer Res.Treat., 62, 51-62.

Wang S.E.; Xiang B.; Guix M.; Olivares M.G.; Parker J.; Chung C.H.; Pandiella A.; Arteaga C.L. (2008). Transforming growth factor beta engages TACE and ErbB3 to activate phosphatidylinositol-3 kinase/Akt in ErbB2-overexpressing breast cancer and desensitizes cells to trastuzumab. Mol.Cell Biol., 28, 5605-5620.

Wang Y.; Thakur A.; Sun Y.; Wu J.; Biliran H.; Bollig A.; Liao D.J. (2007). Synergistic effect of cyclin D1 and c-Myc leads to more aggressive and invasive mammary tumors in severe combined immunodeficient mice. Cancer Res., 67, 3698-3707.

Wang Z.; Wijewickrama G.T.; Peng K.W.; Dietz B.M.; Yuan L.; van Breemen R.B.; Bolton J.L.; Thatcher G.R. (2009). Estrogen Receptor {alpha} Enhances the Rate of Oxidative DNA Damage by Targeting an Equine Estrogen Catechol Metabolite to the Nucleus. J.Biol.Chem., 284, 8633-8642.

Wartenberg M.; Diedershagen H.; Hescheler J.; Sauer H. (1999). Growth stimulation versus induction of cell quiescence by hydrogen peroxide in prostate tumor spheroids is encoded by the duration of the Ca(2+) response. J.Biol.Chem., 274, 27759-27767.

Watanabe R.; Nakamura H.; Masutani H.; Yodoi J. (2010). Anti-oxidative, anti-cancer and anti-inflammatory actions by thioredoxin 1 and thioredoxin-binding protein-2. Pharmacol.Ther.

Weitsman G.E.; Weebadda W.; Ung K.; Murphy L.C. (2009). Reactive oxygen species induce phosphorylation of serine 118 and 167 on estrogen receptor alpha. Breast Cancer Res.Treat., 118, 269-279.

Welsh S.J.; Bellamy W.T.; Briehl M.M.; Powis G. (2002). The redox protein thioredoxin-1 (Trx-1) increases hypoxia-inducible factor 1alpha protein expression: Trx-1 overexpression results in increased vascular endothelial growth factor production and enhanced tumor angiogenesis. Cancer Res., 62, 5089-5095.

Welsh S.J.; Williams R.R.; Birmingham A.; Newman D.J.; Kirkpatrick D.L.; Powis G. (2003). The thioredoxin redox inhibitors 1-methylpropyl 2-imidazolyl disulfide and pleurotin inhibit hypoxia-induced factor 1alpha and vascular endothelial growth factor formation. Mol.Cancer Ther., 2, 235-243.

Wlodek L. (2002). Beneficial and harmful effects of thiols. Pol.J.Pharmacol., 54, 215-223.

Wood Z.A.; Poole L.B.; Hantgan R.R.; Karplus P.A. (2002). Dimers to doughnuts: redox-sensitive oligomerization of 2-cysteine peroxiredoxins. Biochemistry, 41, 5493-5504.

Wood Z.A.; Poole L.B.; Karplus P.A. (2003). Peroxiredoxin evolution and the regulation of hydrogen peroxide signaling. Science, 300, 650-653.

Wood Z.A.; Schroder E.; Robin H.J.; Poole L.B. (2003). Structure, mechanism and regulation of peroxiredoxins. Trends Biochem.Sci., 28, 32-40.

Wu R.F.; Terada L.S. (2009). Ras and Nox: Linked signaling networks? Free Radic.Biol.Med., 47, 1276-1281.

Xiang B.; Chatti K.; Qiu H.; Lakshmi B.; Krasnitz A.; Hicks J.; Yu M.; Miller W.T.; Muthuswamy S.K. (2008). Brk is coamplified with ErbB2 to promote proliferation in breast cancer. Proc.Natl.Acad.Sci.U.S.A, 105, 12463-12468.

Yamashita H.; Nishio M.; Toyama T.; Sugiura H.; Kondo N.; Kobayashi S.; Fujii Y.; Iwase H. (2008). Low phosphorylation of estrogen receptor alpha (ERalpha) serine 118 and high phosphorylation of ERalpha serine 167 improve survival in ER-positive breast cancer. Endocr.Relat Cancer, 15, 755-763.

Yao K.S.; Xanthoudakis S.; Curran T.; O'Dwyer P.J. (1994). Activation of AP-1 and of a nuclear redox factor, Ref-1, in the response of HT29 colon cancer cells to hypoxia. Mol.Cell Biol., 14, 5997-6003.

Yap X.; Tan H.Y.; Huang J.; Lai Y.; Yip G.W.; Tan P.H.; Bay B.H. (2009). Over-expression of metallothionein predicts chemoresistance in breast cancer. J.Pathol., 217, 563-570.

Yeh G.C.; Occhipinti S.J.; Cowan K.H.; Chabner B.A.; Myers C.E. (1987). Adriamycin resistance in human tumor cells associated with marked alteration in the regulation of the hexose monophosphate shunt and its response to oxidant stress. Cancer Res., 47, 5994-5999.

Yu C.X.; Li S.; Whorton A.R. (2005). Redox regulation of PTEN by S-nitrosothiols. Mol.Pharmacol., 68, 847-854.

Yu Q.; Geng Y.; Sicinski P. (2001). Specific protection against breast cancers by cyclin D1 ablation. Nature, 411, 1017-1021.

Zhan L.; Xiang B.; Muthuswamy S.K. (2006). Controlled activation of ErbB1/ErbB2 heterodimers promote invasion of three-dimensional organized epithelia in an ErbB1-dependent manner: implications for progression of ErbB2-overexpressing tumors. Cancer Res., 66, 5201-5208.

In: Oncoproteins: Types and Detection
Editor: Jeremy R. Davis
ISBN: 978-1-61761-551-1

Chapter 2

Regulating c-Myb in Real Time

Anita M. Quintana and Scott A. Ness*
Department of Molecular Genetics and Microbiology, University of New Mexico
Health Sciences Center, Albuquerque, New Mexico, USA

Abstract

The c-Myb transcription factor has been implicated in the control of proliferation and differentiation in many different cell types. While several mechanisms for regulating it have been identified, evidence suggests that the activity of c-Myb can be easily modified and varies in different cell types, and even in a time-dependent manner. For example, c-Myb binds and regulates the promoters of genes that control different parts of the cell cycle, such as CDC2, CCNB1 and CCNA1, suggesting that the activities and specificities of c-Myb, and which genes it regulates, vary during the cell cycle. Wild type c-Myb and its truncated and mutated oncogenic derivative v-Myb regulate distinct sets of genes, suggesting that mutations in c-Myb, which mimic the effects of post-translational modifications, completely change its activity and direct it to different target genes. These findings suggest that protein-protein interactions, subject to the effects of mutations and post-translational modifications, are the predominant modifiers of c-Myb activity, leading to dynamic changes in its specificity in real time. Here we will address the mechanisms that regulate the changes in c-Myb activity in different cell types and during the cell cycle with an emphasis on how these mechanisms go awry in transformed cells.

Introduction

The c-Myb protein is a transcription factor with a highly conserved DNA binding domain near its N-terminus (Figure 1) and several additional conserved motifs comprising a large C-terminal domain involved in negative regulation, control of its degradation and half-life,

* Corresponding author: Email: ness@unm.edu, Department of Molecular Genetics and Microbiology, MSC08 4660, 1 University of New Mexico, Albuquerque, NM 87131-0001 USA

transcriptional activation and the selection of specific target genes [1,2]. The interest in c-Myb stems less from its role as a transcriptional regulator, since it activates genes in a manner similar to many other transcription factors, and more from the dualities of its functions. For example, different versions of c-Myb can be either a critical regulator of normal cells or a potent oncoprotein, and the protein is required for the opposing processes of cell proliferation as well as for terminal differentiation. Although the c-*myb* gene was amongst the first cellular proto-oncogenes to be identified and studied, the activities of the c-Myb protein seem paradoxical and its functions remain enigmatic. Because of the changes in activity and specificity, nomenclature becomes important. For the purposes of this chapter we will distinguish between c-Myb the protein and c-*myb* the gene (MYB) in order to differentiate between these cellular forms and the viral forms v-Myb and v-*myb*.]

I. The Importance of c-Myb in Differentiation, Proliferation and Oncogenesis

The Origin of c-Myb and v-Myb

The c-*myb* gene is the cellular progenitor of the viral oncogene, v-*myb*, which transforms immature hematopoietic cells in culture and induces leukemia in animal models [3-5]. The viral derivative of c-*myb* originated in two chicken retroviruses, Avian Myeloblastosis Virus (AMV) and the E26 virus, each of which encodes a truncated and mutated version of the c-Myb protein [6,7]. Both viruses are potent inducers of acute leukemia in chickens and rapidly transform immature cells by blocking their differentiation and promoting growth factor-dependent proliferation in vitro [8-10]. When the viral oncogenes are expressed from murine retroviruses, they are also capable of inducing leukemia in mice [11,12], and the c-*myb* gene is frequently activated by retrovirus insertions in mouse models of leukemogenesis [13-17]. Even though the v-Myb proteins are potent oncoproteins, the cellular parent, c-Myb does not induce tumors or leukemia in animals, even when constitutively over-expressed [18], and c-Myb can only transform cells in tissue culture in specialized situations [19]. The viral oncoproteins harbor deletions and point mutations and these modifications to the structure of c-Myb greatly enhance transforming activity [20]. The mutations are now known to affect specificity and target gene selection, since c-Myb and AMV v-Myb regulate distinct sets of endogenous genes [21]. This is one of the dualities of c-Myb: that relatively minor changes in its structure can completely change the spectrum of genes that it regulates. The unexpected plasticity in c-Myb activity may be its signature trait, and could explain its relatively underappreciated role as an oncoprotein in human malignancies.

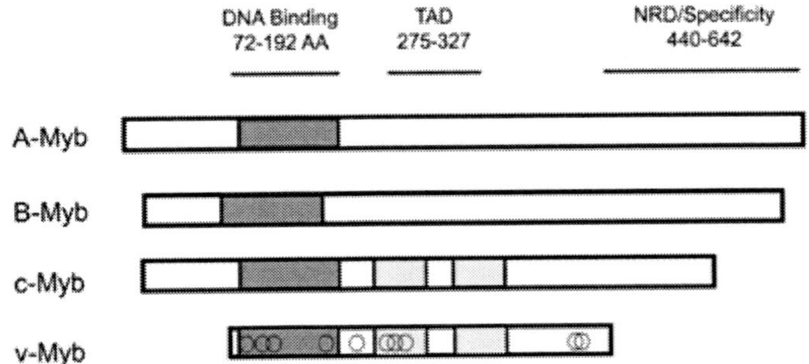

Figure 1. Conserved domains in Myb proteins.

All Myb proteins have a conserved DNA binding domain (shaded dark grey) with a transactivation domain. The large C-terminal domains are known to regulate activity, specificity, and negative regulation. The oncogenic v-Myb is a truncated version of c-Myb, but contains regions of conservation (shaded light grey) and a number of mutations that affect its activity (open circles)

Expression of c-Myb in Human Tumors

Because the v-*myb* oncogenes induce leukemia in animals, most of the attention on c-*myb* as a human oncogene has focused on its role in leukemia and other hematopoietic malignancies. In general, the expression of c-*myb* RNA and c-Myb protein in normal hematopoietic cells correlates with proliferation and decreases as cells terminally differentiate [22]. Thus, the highest levels of c-Myb are found in immature, proliferating cells, and little or no c-Myb is found in mature, differentiated, non-dividing cells [23]. In malignancies, the expression of c-Myb is relatively high in most if not all human leukemias including T-ALL, B-ALL, AML and CML, often at much higher levels than in normal progenitor cells [24-27]. Elevated levels are also detected in several types of solid tumors, most notably colon and breast [28-31]. However, the association between c-*myb* expression and proliferation has led to the question of whether c-*myb* is a causal factor in human tumorigenesis, or merely a marker of higher proliferation in tumors compared to normal cells. Recently, a number of studies suggest a causal role by identifying c-*myb* gene amplifications and/or recurrent rearrangements in a variety of human tumors including leukemia and lymphomas, medulloblastomas, squamous cell carcinomas, small cell lung carcinomas and head and neck tumors [27,32-37]. This evidence strongly suggests that amplification and/or mutation of the c-*myb* gene plays an important and causal role in the development of a wide variety of human hematopoietic, epithelial and neural tumors. Therefore, understanding the functions and activities of the c-*myb* gene and c-Myb protein are critical for understanding a variety of normal processes and for devising treatment strategies targeting this important human oncogene.

Regulation of Hematopoietic Cell Proliferation and Differentiation

The primary role of the c-Myb protein in normal development was established by constructing and analyzing c-*myb*-/- knockout mice, which have normal yolk sac hematopoiesis but succumb due to a defect in definitive erythropoiesis in the fetal liver [38]. Interestingly, conditional alleles that produce reduced levels of c-Myb protein rather than complete eradication have demonstrated that lower levels of c-Myb are sufficient to produce progenitor cells which have a decreased ability to differentiateinto specific lineages [39]. For example, reduced levels (5-10% of wildtype littermates) of c-Myb are sufficient for the production of monocytes, macrophages, and megakaryocytes but impair the differentiation of erythroid cells, suggesting that different steps in the regulation of hematopoiesis require different levels of c-Myb expression. Thus, low levels of expression may suffice for progenitor cell proliferation, but higher levels are needed for differentiation, suggesting that c-Myb plays important roles in the competing processes of differentiation and proliferation.

The c-Myb protein is expressed in all proliferating progenitor cell populations and during differentiation the degree of c-Myb expression generally declines [40-42]. Results demonstrating that constitutive c-Myb expression blocks differentiation [43] and that anti-sense oligonucleotides inhibit proliferation both in vitro and in vivo [44-48] demonstrate that c-Myb is an important regulator of proliferation, at the expense of differentiation. However, the identification of many target genes [1] that are expressed in terminally-differentiated cells suggests that c-Myb also plays a role in stimulating differentiation, which seems contradictory. Understanding this paradox: that c-Myb regulates both differentiation and proliferation, may be the key to understanding the importance and uniqueness of c-Myb as an oncoprotein.

The published results from various laboratories establish that c-Myb has clear biological phenotypes, but do not explain how one protein can have different and apparently contradictory functions. One explanation could be that c-Myb has different functions in different cell types. Most of the studies linking c-Myb to the regulation of proliferation have been performed with hematopoietic cells: usually with leukemia-derived cell lines. For example, c-Myb binds to the CCNA1 (Cyclin A1) and CCNB1 (Cyclin B1) promoters and activates their expression in myeloid leukemia-derived cell lines [49,50]. In these cells, where c-Myb is highly expressed, it could play an important role in promoting progression through the cell cycle, which could in turn be regulated by interactions with Cyclin D1 [51] and Cyclin Dependent Kinases 4 and 6 [52]. The interactions between Cyclin D1 are different for c-Myb and v-Myb, raising the intriguing possibility that regulation of the cell cycle is an integral part of the oncogenic potential of c-Myb [51]. It is not yet clear whether c-Myb also regulates the expression of the Cyclin genes in normal cells, or in only a subset of normal cells such as hematopoietic cell progenitors, although there is some evidence that c-Myb may regulate the Cyclin E1 gene in some proliferating epithelial cells [53] and that c-Myb plays an important role in the estrogen-induced proliferation of mammary cells [54], suggesting that the link between c-Myb and proliferation may be more general.

There is also ample evidence that c-Myb plays an important role in hematopoietic cell differentiation. The original c-Myb target gene, *mim-1*, is only expressed in differentiating neutrophils [55], and this specificity is linked to the requirement for combinatorial cooperation with NF-M, the chicken equivalent of C/EBPbeta [56]. This is a model for the lineage specificity of other c-Myb target genes, which likely require that c-Myb cooperates with other regulators. Such targets include lineage specific genes like neutrophil elastase [57,58] and myeloperoxidase [59], as well as target genes that are expressed only in immature cells such as CD34 [60] and c-KIT [61,62].

Alternatively, c-Myb activity could be qualitatively different in different cell types because of lineage-specific post-translational modifications, such as serine or threonine phosphorylation, arginine methylation, lysine acetylation or others [2]. The post-translational modifications could affect the specificity and selection of target genes by c-Myb, analogous to the way point mutations in v-Myb change its activity. These possibilities and the effects of post-translational modifications on c-Myb activity will be discussed below.

Different Levels of c-Myb May Lead to Different Outcomes

In addition to qualitative changes in c-Myb activity, or differences in which cooperating factors are available in different lineages, there is also evidence that quantitative changes in c-Myb levels lead to the activation of different pathways. As discussed previously, in mouse models low levels of expression favor the formation of macrophages and monocytes, whereas higher levels favor erythropoiesis [39]. In general, the level of c-Myb activity in different cell types is difficult to assess, especially since c-*myb* RNA levels decrease during differentiation [41-43,63]. How is it possible that ablation of c-*myb* is lethal, but small changes in expression dictate lineage selection? While it is true that c-Myb is essential, a threshold of expression may exist that modulates proper hematopoietic cell differentiation. For example, despite the fact that c-*myb* RNA levels decrease substantially during lymphoid cell differentiation, c-Myb protein binds to and activates the promoters of genes that are expressed in fully differentiated cells such as the VDJ recombinase RAG2 [64,65]. RAG2 is only expressed in fully differentiated T and B-cells, which express little if any c-*myb* mRNA, yet c-Myb protein can cause activation of this gene during an immune response. Results such as these suggest that the level of c-*myb* RNA makes a poor predictor of c-Myb protein activity. Therefore, it seems likely a major part of c-Myb protein regulation occurs through post-translational changes such as stability, modifications and changes in protein interactions that can be performed and reversed quickly, depending on cell context, and using residual c-Myb protein, even in cells that express little or no c-*myb* mRNA.

II. Context-Specific Changes in c-Myb Activity

Structural and Conserved Domains

The dominant conserved feature in the c-Myb protein is the DNA binding domain near the N-terminus, which is identical in c-Myb proteins from species as diverse as humans, mice and chickens (Figure 1). The c-Myb DNA binding domain has been used as a signature to identify two additional transcription factors, A-Myb (MYBL1) and B-Myb (MYBL2), which have nearly identical DNA binding domains and similar overall structures, but very different biology and functions. Both crystal and solution structures of the DNA binding domain, either free in solution or bound to DNA, have been determined (reviewed in [1]). Interestingly, even the parts of the DNA binding domain that do not contact DNA have been perfectly conserved amongst vertebrate c-Myb proteins, suggesting that the DNA binding domain also serves as an important protein binding or docking surface [1]. Indeed, several proteins have been identified that interact with the c-Myb or v-Myb DNA binding domains [51,66-68]. More importantly, three of the mutations in the AMV v-Myb protein change amino acids mapping to protein-interaction surfaces in the DNA binding domain [1]. These amino acid changes alter the activity of the protein [56], change the phenotypes of transformed cells [69] and shift the activity on some target genes [70], showing that relatively subtle changes in the structure of c-Myb, even affecting just a single amino acid residue, can change which target genes get regulated.

There are numerous regions within the c-Myb protein that are highly conserved and have been shown to have very important functions. The conserved trans-activation domain, amino acids 275-327, is essential for the transcriptional activity of c-Myb and v-Myb as measured in reporter gene assays [71,72]. The extreme C-terminal domains negatively regulate the protein and truncation or deletion of this region causes increased c-Myb activity in some assays [66,73]. The EVES motif, near the C-terminus and centered at amino acid 530, is involved in negative regulation by inducing an intramolecular fold-back mechanism and interacting with the DNA binding domain [66]. As discussed further below, the c-Myb protein has many activities that contribute to its ability to regulate genes and control proliferation and differentiation.

Myb Proteins Have Chromatin Remodeling Activities

Although the Myb DNA binding domain defines a family of related proteins, it recently acquired the additional classifier as a SANT domain protein, defined by a conserved motif found in multiple chromatin remodeling enzymes with homology to the Myb DNA binding domain [74]. The SANT domain is named for 'switching-defective protein 3 (Swi3), adaptor 2 (Ada2), nuclear receptor co-repressor (N-CoR) and transcription factor (TF) IIIB [75]. The structural link between c-Myb and these well-known chromatin-remodeling enzymes suggests that c-Myb could also play a role in chromatin remodeling. The DNA binding domain of c-Myb has been shown to bind and facilitate the acetylation of the N-terminal tails of histones H3 and H3.3, and this interaction is disrupted by the v-Myb mutations in the DNA binding domain [76], suggesting that changes in the chromatin remodeling activities of c-Myb are important for unleashing its oncogenic activity. Fine mapping of the conserved residues within the c-Myb DNA binding domain also led to the identification of an acidic patch required for interactions with the N-terminal tails of histone H4 [77], suggesting that the c-Myb DNA binding/SANT domain may have multiple histone binding activities that are part of its chromatin remodeling function. However, c-Myb is not known to have an acetyltransferase domain, so it must facilitate changes in chromatin structure by mediating or directing the activities of histone modifying enzymes.

In addition to direct interactions with histone tails, c-Myb can also direct chromatin remodeling by interacting with enzymes that modify histones. For example, c-Myb is found in a complex with the Mixed-Lineage Leukemia protein (MLL), a SET domain-containing histone methyltransferase that is frequently a component of fusion proteins produced by chromosome translocations in human acute leukemia. MLL methylates lysine 4 of histone H3 (H3K4), which is an important step in gene activation. A third protein, Menin, an MLL interacting partner [78], mediates the interaction between MLL and c-Myb. Thus, the gene activation activity of MLL may be directed to specific target genes through MLL-Menin-Myb complex formation. Since MLL is a known human oncogene, this association suggests a possible mechanism of c-Myb in leukemogenesis.

The Myb DNA binding domain recognizes sites in DNA matching a loose consensus sequence: (C/T)AAC(G/T)G [79]. Because this sequence is degenerate, it is expected to occur on average about once per kilobase, or more than 3 million times in the haploid human genome. Clearly, other domains in the c-Myb protein, which likely facilitate specific protein-protein interactions, must also play an important role in determining which binding sites are preferred and therefore which genes get regulated. This is illustrated best by microarray experiments that identified hundreds of target genes that were activated or repressed when different Myb proteins were ectopically expressed either in breast cancer cell lines or in primary human monocytes. To a first approximation, despite having nearly identical DNA binding domains, c-Myb activated totally different sets of genes than either A-Myb or B-Myb [80] and completely different genes than were activated by its oncogenic derivative, v-Myb [21]. Furthermore, domain swap experiments showed that the DNA binding domains were essentially interchangeable and that the biggest effects on target gene selection were caused by the other parts of the protein [70]. Thus, the regions of c-Myb outside the DNA binding domain, especially the large C-terminal domains that are often referred to as the transcriptional activation and/or negative regulatory domains, are in fact deeply involved in the selection of target genes and in determining the specificities of the c-Myb protein in different situations.

Myb-like Domains in Other Proteins

In addition to chromatin remodeling proteins, Myb-like DNA binding domains are found in the important regulator DMP1, a Cyclin D1-binding transcriptional regulator that has tumor suppressor activity [81,82], as well as in a group of important regulators called the Telebox proteins [83]. The Telebox domain is found in the human TRF1 and TRF2 proteins and is responsible for the ability of these proteins to bind telomere repeats in DNA after forming homodimers [84]. This mechanism sets the Telebox proteins apart from Myb proteins, since the latter do not form homodimers. Although the Myb-like domains are important for telomere binding, the Telebox proteins can bind to non-telomeric DNA as well, suggesting that these domains are specialized and may also have other functions within the cell [85]. Interestingly, TRF1 and c-Myb also share the ability to interact with the Pin1 proline isomerase, which regulates protein conformations in a phosphorylation-dependent manner [86,87]. The c-Myb protein also interacts with the proline isomerase Cyp40, which induces conformational changes in c-Myb but not v-Myb [67], suggesting that interactions with enzymes like Cyp40 and Pin1 could play an important role in suppressing or regulating the oncogenic activity of c-Myb.

Post-Translational Modifications of c-Myb

The c-Myb protein is subject to numerous post-translational modifications that can affect its activity (Figure 2A). Phosphorylation by Casein Kinase II at the N-terminus of c-Myb has been linked to changes in DNA binding activity and tumorigenic potential, since the phosphorylation sites are absent in the oncogenic allele, v-Myb [88,89]. At least two kinases, Protein Kinase A and Pim-1, can phosphorylate the DNA binding domain of c-Myb [68,90]. Interestingly, the recognition site for the former is disrupted by one of the oncogenic point mutations in v-Myb, suggesting that loss of recognition by Protein Kinase A could be a factor in oncogenic activity, or in altering the specificity and target gene recognition by v-Myb [90]. The c-Myb protein is also phosphorylated at several sites in the C-terminal domains, and phosphorylation has been implicated in negative regulation, control of intramolecular interactions and changes in transcriptional activity [66,91-100]

Phosphorylation of c-Myb has also been implicated in regulating subsequent post-translational modifications. SUMOylation is the reversible process of covalent modification of proteins with small ubiquitin like modifiers (SUMO). There are four SUMO proteins and c-Myb is SUMOylated at several different residues in response to different cellular signals. SUMO modification of c-Myb has been divided into distinct classes. The first is SUMOylation by SUMO protein 1 [101,102] and occurs in the negative regulatory domain through an interaction with the SUMO modifying enzyme UBC9 [101]. This modification is highly dependent upon the presence of lysine 523 and results in decreased c-Myb transactivation activity. Although SUMOylation of c-Myb by SUMO 2/3 occurs at the same sites, it is added as a response to cellular stress and requires a separate E3 ubiquitin ligase [103]. SUMO 2/3 modifications cause a reduction in the activation of c-Myb target genes [103] and are dependent on the interactions of c-Myb with either UBC9 or PIASy. These examples show that changes in protein interactions and post-translational modifications, even those that occur outside of the DNA binding domain, can have a dramatic effect on c-Myb activity.

Phosphorylation has also been linked to ubiquitinylation and subsequent degradation of c-Myb. Turnover of c-Myb, which is mediated by ubiquitinylation followed by degradation by the 26S proteasome, is more rapid in normal cells and less efficient in tumors, and mutation and deletion of specific regions/residues leads to increased stability [104]. The E3 ubiquitin ligase and F-box protein Fbw7 interacts with and ubiquitinylates c-Myb, stimulating its degradation [96]. This first requires phosphorylation of threonine residue 572, a substrate for Glycogen Synthase Kinase 3 [96]. Wnt pathway signaling has been shown to stimulate kinases TAK1 (TGF-beta-activated kinase), HIPK2 (homeodomain-interacting protein kinase 2), and NLK (Nemo-like kinase), all of which can phosphorylate sites in the C-terminal domains of c-Myb and stimulate its subsequent ubiquitinylation and proteasome-dependent degradation [95]. Regulation by Wnt pathways suggests that phosphorylation and degradation of c-Myb is regulated by signals that stimulate the differentiation of epithelial and other cell types, suggesting that the levels of c-Myb play a critical role in keeping such cells immature, or in stimulating their proliferation.

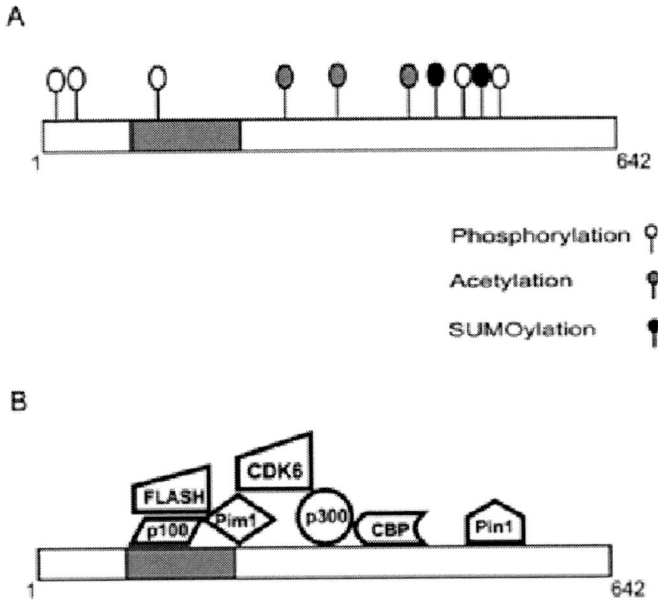

Figure 2. c-Myb is actively modified after translation. A. c-Myb is phosphorylated by at least seven different protein kinases and the C-terminal domains are SUMOylated, ubiquitinylated, acetylated. These modifications affect protein interactions, stability, and activity. B. c-Myb interacts with different types of proteins and many of them binding to the DNA binding domain. A select few such as Pim1 and p300 result in post-translational modifications that affect activity.

Regulation of c-Myb Activity by Acetylation

c-Myb interacts with numerous co-activators and in some cases, these interactions affect Myb activity. Mutation of c-Myb can affect these interactions. One such mutation is M303V, which disrupts an interaction with the transcriptional coactivator and histone acetyltransferase, p300. In a mouse model system, this disruption leads to defects in hematopoiesis and promotes cellular proliferation [105] suggesting that the interaction between c-Myb and p300 affects cell fate decisions. p300 acetylates c-Myb at three lysines within the C-terminal domain and mutation of these residues to alanine leads to an increased DNA binding ability and increased transactivation activity [106]. Moreover, c-Myb interacts with the p300-related CREB Binding Protein (CBP) near the c-Myb transactivation domain, resulting in increased transactivation activity [107,108]. CBP and p300 can acetylate the same residues in c-Myb, suggesting that these two proteins can have redundant functions, but it is unclear which interacting partner is preferred during different developmental stages or in different growth conditions. For example, CBP can interact with the transactivation domain or the negative regulatory domain [108], this difference could account for preferences in which residue gets modified, which in turn might reflect differences in the chronological sequence of protein modifications that dictate c-Myb specificity. If there is a context specific code [2] mediating c-Myb activity then the sequence of modifications could lead to changes in c-Myb specificity.

One of the problems with many of the experiments that have measured the impact of post-translational modifications is that they have only used relatively crude measures of c-Myb activity. Since the cloning of the original Myb-regulated target gene, mim-1, it has been clear that the ability of c-Myb and v-Myb to activate endogenous, chromatin-embedded target genes does not correlate with the ability to activate plasmid-based reporter gene assays [55,56,69]. With the advent of microarray assays, the complexities of target gene selection by different versions of c-Myb have become evident [21,80], and it seems likely that phosphorylation, acetylation, SUMOylation and other post-translational modifications may be responsible for shifting or determining the selection of target genes by c-Myb. To realize the full potential of these modifications, it will be necessary to test different versions of c-Myb in more sensitive, unbiased gene activation assays to determine how these modifications affect c-Myb in the context of the cell and in real time.

Implications of Post-Translational Modifications: Subpopulations of c-Myb with Different Activities

As discussed above, at least 7 protein kinases have been identified that can phosphorylate c-Myb under a variety of different conditions, in response to different types of stimuli, and at various locations throughout the length of the protein. In addition, several protein phosphatases probably play an important role in regulating c-Myb activity by countering the effects of the kinases [94], and the phosphorylation status regulates the interactions with several E3 ubiquitin ligases or sumo ligases that make further modifications, potentially affecting degradation. Like many transcription factors, the c-Myb protein is also acetylated by transcriptional co-activators CBP and p300 [106,108], and it is likely to have additional modifications that have not yet been characterized. Altogether, there are at least 10 identified enzymes that make post-translational modifications to c-Myb, which are likely to affect stability, activity, interactions with other proteins and target gene selection. Clearly, post-translational modifications are an important source of c-Myb regulation.

Myb interacts with a wide array of proteins (Figure 2B). Some protein interactions may be dependent on or lead to new post-translational modifications. In addition, post-translational modifications may occur in a chronological order, analogous to the way modifications are added sequentially to histones during transcriptional activation. These changes in post-translational modifications could account for the ability of c-Myb to bind to and regulate different genes in different biological processes. As discussed, c-Myb regulates the CCNA1, CCNB1, and CCNE1 genes [50,53,109] during the cell cycle. In the time-dependent cell cycle, c-Myb could be bound to these genes all the time but only participate in regulating the genes in specific stages of the cell cycle. Alternatively, c-Myb binding to specific promoters could change in a time-dependent manner, oscillating between different promoters as the cells progress through the cell cycle. To distinguish between these possibilities, it will be necessary to measure the association of c-Myb with different target genes as cells progress through the cell cycle.

The idea that c-Myb activity and target gene selection could change in a time-dependent manner during the cell cycle runs counter to expectations of transcription factor activities based solely on plasmid-based reporter gene assays. This may be the reason that the ability of c-Myb to activate endogenous genes does not correlate with the ability to activate reporter genes. If post-translational modifications affect target gene selection by c-Myb, it is likely that the cell contains multiple pools of c-Myb with different, perhaps overlapping specificities. For example, phosphorylation of c-Myb is likely to be sub-stoichiometric, defining a subset or pool of protein with a unique post-translational modification and perhaps a unique specificity for select target genes. However, modification of c-Myb is not limited to phosphorylation since the protein can also be acetylated [106,108], SUMOylated[101-103], and ubiquitinated [110], and is likely also to be modified in other ways. Potentially, each modification could distinguish a specific sub-population of c-Myb proteins with a unique spectrum of activities, directed toward a specialized subset of target genes by unique protein-protein interactions that are controlled by the modifications.

Regulating transcription by post-translational modifications or protein interactions offers the ability of regulating activity in a dynamic, time-dependent manner since these types of modifications can occur rapidly in response to mitogenic or differentiation signals. There is very little known about how c-Myb activity might be affected by upstream signaling pathways, such as cytokine receptor pathways or oncogenic tyrosine kinases. Now that assays are available to follow subtle changes in c-Myb target gene selection, it will be interesting to see how such signaling pathways might affect c-Myb activity and its choice of target genes in specific situations in normal cells and in malignancies.

III. Mechanisms that Control c-Myb Expression, Activity and Specificity

Alternative RNA Splicing Produces Multiple forms of c-Myb

As discussed above, subpopulations of c-Myb within a cell can be generated through sub-stoichiometric post-translational modifications. Another mechanism that produces multiple forms of c-Myb is alternative RNA splicing. The c-*myb* gene contains 15 exons and undergoes alternative splicing in both cancer cells and normal cells [111-114]. The vast majority of alternative splicing occurs in exons 6-15 and excludes the exons that encode the DNA binding domain. Thus, almost all of the variant c-Myb proteins encoded by alternatively spliced mRNAs contain the same DNA binding domain, but have changes in the large C-terminal domains controlling transcriptional activation, negative regulation and target gene specificity [112]. For example, the addition of alternative exons such as exon 6A, which results in a 400 bp insertion between exon 6 and 7, have been described in normal cells and cancer cells [113]. Exon 9 can form up to four different variants in both chickens and humans and at least two of them are found in hematopoietic cells [112,115,116]. With the identification of additional c-Myb target genes and the availability of assays that can detect differences in target gene selection by different versions of c-Myb, it has become clear that many of the proteins encoded by splice variants have distinct transcriptional activities and activate different sets of genes than the "wild type" c-Myb [112]. These results greatly increase the amount of complexity that is possible in the targeting of genes by c-Myb, since different subpopulations of the protein, encoded by different versions of the mRNA, could regulate entirely different sets of target genes.

Comparisons of different types of cells have shown that different patterns of c-*myb* mRNA splicing are found in different cell lineages [112], suggesting that alternative mRNA splicing is as carefully regulated as any other aspect of differentiation. In addition, much higher levels of c-myb alternative splicing are found in leukemias than in normal bone marrow cells [112], suggesting that the mechanisms controlling alternative splicing are defective or modified in malignancies. Thus, amplified alternative splicing may be a novel mechanism for activating oncogenes like c-myb, which are capable of producing variant proteins with different, perhaps transforming, activities. This idea is strengthened by recent findings showing that regulators of alternative RNA splicing are themselves oncogenes [117,118], presumably because they cause the production of alternatively-spliced mRNAs that encode proteins with altered/oncogenic activities, similar to the altered activities displayed by v-Myb compared to c-Myb.

One of the major questions that remains is, do different versions of c-Myb, encoded by different splice variants or produced by sub-stoichiometric post-translational modifications, compete for the same target gene promoters? For example, phosphorylation of specific residues in the N-terminus of c-Myb may only occur on a fraction (1%) of the total molecules of c-Myb. These phosphorylated molecules may have different activities and different binding specificities, due to different cooperative interactions with other proteins that bind some target promoters. If subpopulations of c-Myb are able to bind specific subsets of target promoters, without really competing with the vast bulk of the protein that has a different specificity, then even relatively rare forms of the protein could make important contributions to the patterns of genes that get expressed, and could contribute to changes in phenotype or to transformation. Additional experiments will be required to determine whether different versions of c-Myb, which appear to have distinct target gene specificities, are able to regulate their target genes in the presence of other, more abundant versions of the protein.

Regulation of c-Myb by MicroRNAs

MicroRNAs are small non-coding RNAs approximately 22 nucleotides in length that interact with and regulate the stability of other RNA transcripts [119,120]. These non-coding RNAs are derived from primordial RNAs that are cleaved into one or more microRNAs complementary to regions within the 3' untranslated region of specific target mRNAs. Genes that encode microRNAs such as lin-4 and let-7 were discovered first in C. elegans and play important roles in the regulation of larval development [120]. In mammals, deregulated expression of microRNAs can influence a large number of genes and can contribute to many different types of cancer including breast and oral cancer [121-124].

The c-*myb* 3' untranslated region (UTR) is extensive and contains many potential binding sites for microRNAs. miR150 is expressed in resting T and B-cells, but not proliferating progenitors [125] and deletion of miR150 can cause an increase in c-*myb* expression and expansion of the B1 subset of B-cells. These observations were more pronounced in B-cells than T-cells presumably because the level of miR150 in T-cells is reduced relative to B-cells [126]. The regulatory relationship between miR150 and c-*myb* is conserved in zebrafish suggesting that this ancient regulatory mechanism has been conserved throughout vertebrate evolution [127]. The miR150 mechanism also regulates c-*myb* expression in other cell types.

For example, expression of miR150 in megakaryocyte-erythrocyte progenitors leads to a profound bias of differentiation towards megakaryocytes, and if c-myb is expressed without the 3'UTR there is a decrease in the formation of megakaryocytes [128]. This is consistent with results showing that a knockdown of c-Myb expression results in increased megakaryocyte production [129]. These data suggest that miR150 is a potent regulator of c-*myb* and differentiation, however other microRNAs have been demonstrated to be of equal importance in regulating c-*myb*.

MiR15a and miR-16 are encoded by the dleu2 tumor suppressor locus, which is regulated by the Pax5 transcription factor, a c-Myb interacting partner [130]. During levels of high Pax5 activity miR15a/16 are down regulated resulting in increased c-Myb protein expression. Likewise, high levels of the microRNAs decrease c-Myb and A-Myb expression, suggesting that some microRNAs regulate multiple Myb family members [130]. Further analysis has established that the c-Myb protein binds the promoter for the gene encoding miR15a/16 and that functional miR15a can inhibit erythroid and myeloid colony formation [131]. These results suggest that one of the ways that c-Myb controls hematopoiesis, and its own expression, is by regulating the expression of microRNA genes. Other microRNAs such as miR34a are complementary to the c-*myb* 3' UTR and recent studies have demonstrated that miR34a regulates c-Myb expression during megakaryocyte development [132], suggesting that different microRNAs can promote different cellular processes.

Abnormal regulation of microRNAs has been implicated in numerous forms of cancer. For example, miR15a is deleted or under-expressed in many tumors. Since many miR15a targets are important for cell cycle regulation [133], these results suggest that loss of miR15a leads to changes in the cell cycle. In contrast, other microRNAs such as miR30 are up-regulated in subsets of cancers [134]. Thus, the regulatory pathways leading to microRNAs are complex and deregulation can occur in the form of deletion, amplification, mutation, and silencing of the genes via epigenetic marks [135]. Since c-Myb is predicted to bind and regulate many microRNA genes [136], it could play an important indirect role in many of processes that are subsequently controlled by microRNAs.

Attenuation as a Regulator of c-*myb* Transcription

There is differential expression of c-*myb* in a variety of different cells during differentiation and changes in c-*myb* expression can be attributed to attenuation of transcriptional elongation in B-cells [137]. This attenuation is mediated by sequences within the first intron of the c-*myb* gene, which includes a DNase I hypersensitive region [138]. During differentiation there is a reduction of protein binding to intron I, correlating with decreased c-*myb* expression [139]. The intron I region contains binding sites for a number of transcription factors including AP-1, SP1, and the estrogen regulated P2 protein [139], and in breast cells the attenuation is relieved in the presence of estrogen [54]. This suggests that intron I attenuation is regulated by mitotic signals in a cell cycle dependent manner. Mutation of the intron I region occurs in some colon cancers, which could lead to the over-expression, contributing to transformation [140]. Thus, a number of transcriptional and post-transcriptional mechanisms are involved in controlling c-Myb expression, and these mechanisms play important roles in controlling differentiation and proliferation.

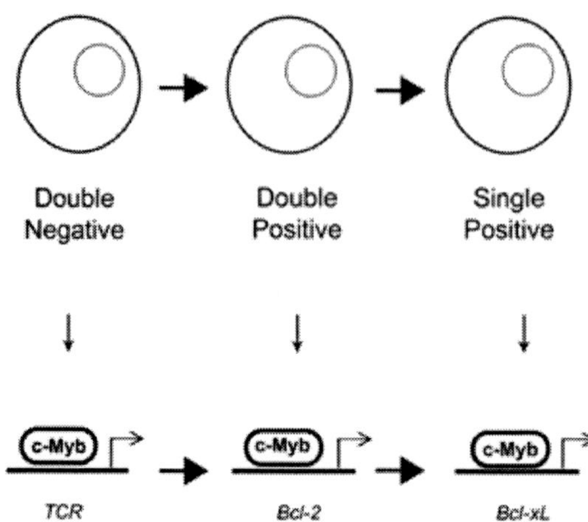

Figure 3. c-Myb sub-populations regulate T-cell development.c-Myb is essential for proper T-cell development and defects in c-Myb expression cause different phenotypes in different stages of development. c-Myb has a wide array of target genes and it is likely that different versions of c-Myb have different activities and specificities in different stages of development and in different cell types.

Mechanisms Regulating c-Myb in T-Cell and B-Cell Development

Although the original c-*myb* knock out animals demonstrated the requirement for c-Myb expression in erythropoiesis, more recent experiments using conditional knock out strains have shown that c-Myb is also necessary for the development of the lymphoid lineages. Conditional knockdowns suggest that c-Myb is critical for three points of thymocyte development: double negative, double positive, and single positive [141]. Knockdown of c-Myb has different effects on each stage of development. For example, increases in double negative thymocytes (DN) upon c-Myb knockdown are not due to defects in the cell cycle or cell survival. In contrast defects in the double positive (DP) T-cells to single positive (SP) cells are due to defects in cell survival [141]. DN cells go through stages of DN(1-4) development and it was demonstrated that at the DN3 stage, levels of c-Myb target gene TCRß were decreased in the c-*myb* knockout cells, resulting in defects in differentiation [141]. These data suggest that as specific cell types (i.e. different T-cell stages) progress through differentiation the same protein, c-Myb, can facilitate multiple processes (Figure 3). It is likely that c-Myb, or more specifically a unique subset of c-Myb proteins, binds to and regulates the TCR during DN3 development, but a different subset of c-Myb regulates other genes to affect cell survival. In theory, this mechanism would affect proliferation by regulating target genes such as cdc2 [142] in a cell cycle dependent manner. c-Myb is essential for proper proliferation and regulation of genes such as cdc2, but can also regulate genes such as Bcl-2, an anti-apoptotic gene [143,144], suggesting that different sub-populations of c-Myb can regulate different genes in different cell types, or even the same lineage of cells at different stages of differentiation. The c-Myb proteins in these sub-populations of cells are likely defined by different post-translational modifications and protein interactions that affect the target gene selection by c-Myb. Thus, knockdown of c-Myb in T-cells has identified different c-Myb subpopulations, each of which is essential for a key stage in cellular development and progression. Recently other groups have expanded on the idea that c-Myb affects different subsets of thymocytes and have demonstrated that during progression from DP to SP c-Myb regulates the anti-apoptotic gene Bcl-xL. Thus, DP thymocytes with c-Myb depleted undergo apoptosis leading to a decrease in specific subsets of SP cells [145]. This concept of sub-populations of c-Myb regulating discrete stages of differentiation can be expanded to other cell types, in particular B-cells, where c-Myb regulates the survival of a specific subset of B-cells [146].

The role of c-Myb in T-cell and B-cell development is complex and it is clear that c-Myb target genes vary in different subpopulations of cells. More specifically, the activity of c-Myb varies in different stages of differentiation. This raises an interesting question: If only a subset of c-Myb molecules is involved in regulating specific genes, what is the rest of the c-Myb doing? Is the cell cycle specific fraction of c-Myb expressed, but not active during the cellular response to differentiation cues? Where does it go if its not bound to genes? Some c-Myb interacting partners have variable localization and could sequester c-Myb protein. For example, the interaction between c-Myb and FLASH leads to colocalization into nuclear speckles, however only a fraction of c-Myb is found in these speckles [147]. Unfortunately, it is not clear how many subpopulations of c-Myb exist within a given cell type, where those subpopulations are localized, how sub-nuclear localization of c-Myb affects which target genes get regulated, or how such regulation changes in real time.

IV. Implications for the Regulation of c-Myb during the Cell Cycle

Myb Genes are Cell Cycle Regulated

Stimulation of primary peripheral blood lymphocytes (PBLs) has demonstrated that c-*myb* is expressed in a cell cycle dependent manner. c-*myb* RNA levels peaks in expression during the G1/S transition [148,149] and protein expression can be detected slightly later [150]. These changes in expression level suggest that cell cycle specific mechanisms participate in the induction of c-Myb during different phases of the cell cycle. The c-*myb* gene promoter contains binding sites for the cell cycle transcription factor E2F [151], which could be one mechanism for explaining the induction of the c-*myb* gene during the cell cycle.

The two related proteins, A-Myb and B-Myb, have distinct patterns of expression during the cell cycle. In hematopoietic cells, expression of B-Myb peaks during the G1/S transition and mutation of the E2F binding site in the B-Myb promoter leads to defective B-Myb gene induction [152]. Both c-*myb* and MYBL2 (B-Myb) appear to peak in expression during G1/S, but expression of the A-Myb gene (MYBL1) peaks in G0/quiescent hematopoietic cells [22], showing that each of the genes is regulated differently. Interestingly, some cell type specificity exists, since in vascular smooth muscle cells A-Myb transcription peaks at the G1/S transition [153]. These data establish that all three genes are cell cycle regulated, although the patterns of regulation may vary in different cell types.

Regulation of the Cell Cycle by Myb Proteins

Several types of evidence have linked the Myb transcription factors to cell cycle regulation. B-Myb is the most closely related to the Drosophila DMYB gene [154]. Mutated alleles of DMYB have detrimental effects on the cell cycle [155] and lead to genomic instability [156]. Recent studies have shown that DMYB is part of the dREAM complex that coordinates specific cell cycle regulation of target genes, particularly in G2/M [157,158]. B-Myb, but not c-Myb, is an active component of the Mip/Lin-9 complex, the mammalian ortholog to the dREAM complex. As part of the Mip/Lin-9 complex, B-Myb regulates genes in a G2/M specific manner. B-Myb is also required for proper regulation of the mitotic spindle [159-162] and for inner cell mass formation [163]. These activities appear to be specific for B-Myb, highlighting the importance of B-Myb in both cell cycle regulation and differentiation, and demonstrating that different Myb proteins have completely different functions.

The A-Myb protein is also associated with proper cell cycle progression. Knockdown of A-Myb leads to an arrest of spermatogenesis and defects in breast cell development [164,165]. These tissues undergo a high level of proliferation suggesting that A-Myb is essential for proliferation in these tissues. Neither A-Myb nor B-Myb are established oncogenes, although expression of each has been detected in B-cell lymphoma or breast cancer, respectively [166,167]. The relative over-expression of A-Myb and B-Myb may reflect the higher rates of proliferation of the tumor cells compared to the normal tissues.

The importance of c-Myb to cell cycle regulation is less well characterized, but it may be regulated in a way that is analogous to that of B-Myb. The genes encoding several important cell cycle regulators, including Cyclins A1, B1 and E1, are regulated by c-Myb, at least in some cell types [49,50,53]. However, these genes are induced in different phases of the cell cycle, so if c-Myb is involved in their regulation, then c-Myb activity or specificity must also be regulated by the cell cycle. Since c-Myb interacts with Cyclin D1 [51] and with cyclin dependent kinases CDK4 and CDK6 [52], it is possible that c-Myb specificity is regulated through a mechanism of cell cycle stage-specific phosphorylation, similar to the regulation of the Retinoblastoma protein and the E2F transcription factors [168,169]. However, cell cycle specific phosphorylation of c-Myb has not yet been documented. Alternatively, c-Myb may only be an accessory transcription factor that cooperates with other proteins that are more specifically regulated. For example, c-Myb could form cooperative complexes with E2F at specific promoters, in the way it does with C/EBPbeta. The direct regulation of E2F could recruit c-Myb to specific promoters at different stages of the cell cycle, making c-Myb more of a passive participant, rather than an active component of cell cycle regulation.

Conclusion

Although the c-*myb* oncogene was one of the first cellular proto-oncogenes to be identified [170,171], the functions of the c-Myb transcription factor that it encodes have remained obscure. In hindsight, part of the problem has been the use of inappropriate measures of c-Myb activity, such as promoter-reporter gene assays, which only assess quantitative changes in c-Myb activity and cannot address the qualitative changes that permit c-Myb to regulate different genes in different situations and in response to variable stimuli [21].

It is likely that c-Myb is regulated by mechanisms that are similar to the cell cycle regulated transcription factors p53 and E2F. The mechanism of p53 cell cycle control is mediated mostly by MDM2, which facilitates proper degradation of p53 [172]. A protein counterpart of MDM2 for c-Myb has not been discovered, but protein interactions can affect the activity of the protein. On the other hand, there is an entire family of E2F proteins that can affect both proliferation and apoptosis, which at times appears contradictory [173]. This type of antagonism is exactly what c-Myb appears to have during proliferation and differentiation. Although there are fewer c-Myb family members, alternatively spliced isoforms of c-Myb, with distinct transcriptional activities, may fulfill those roles. Different post-translationally modified subsets of c-Myb have the potential to mediate the proliferation or differentiation of cells, in response to different signaling pathways. Regulation of other proteins could add some insight about c-Myb activity during proliferation and differentiation. Therefore, it is critical to consider the mechanism of other cell cycle regulated transcription factors.

The most highly related transcription factors to compare c-Myb with are A-Myb and B-Myb. A-Myb and B-Myb interact with cyclin A: CDK2, which leads to phosphorylation of each protein [109,174]. Phosphorylation of A-Myb leads to a relief of transcriptional inhibition and phosphorylation of B-Myb increases transactivation. These results with Myb related partners demonstrate two important results. First, the same complex interacts with and phosphorylates each Myb protein, which has distinct affects on Myb activity. Second, cell cycle machinery such as cyclin A: CDK2 can interact with multiple Myb proteins. This suggests Myb proteins, including c-Myb, may shuffle from one multiprotein complex to another, regulating different sets of promoters in different stages of the cell cycle (Figure 4). This could affect specificity and explain how c-Myb can activate both the CCNB1 and CCNA1 genes. In addition, it is plausible that this shuffling is coordinated amongst Myb proteins. As described above, B-Myb activates G2/M specific genes as a component of the dREAM complex, but c-Myb can bind to some of the same promoters. In addition c-Myb interacts with CCND1: CDK6 suggesting that it may participate in complexes similar to the dREAM complex to regulate different sets of genes, or the same genes as B-Myb at different stages of the cell cycle (Figure 5).

This chapter has described the hypothesis that c-Myb post-translational modifications and protein interactions account for changes in c-Myb activity. Results from many laboratories have documented the extensive regulation of c-Myb by post-translational modifications that affect its stability, activity and specificity towards target genes. A disparity between c-Myb activity and c-*myb* expression exists and results suggest that cells may have several sub-populations of c-Myb, defined by different post-translational modifications or produced through alternative RNA splicing, that simultaneously regulate unique sets of target genes. These sub-populations are presumably formed in response to signaling pathways activated by cell surface receptors, leading to changes in post-translational modifications, alternative RNA splicing and the formation of alternative multiprotein complexes containing c-Myb that interact with and regulate different target promoters. Future experiments will need to define these sub-populations and dissect the mechanisms that allow c-Myb to regulate so many different genes in different situations, and to understand how these mechanisms go awry when c-Myb is transformed into an oncoprotein.

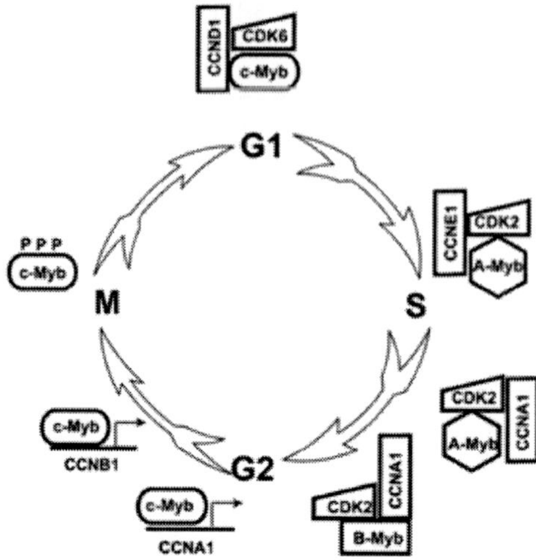

Figure 4. Myb proteins are cell cycle regulated. Components of the cell cycle such as cyclin: CDK complexes can interact with all Myb proteins. A-Myb and B-Myb are phosphorylated by different pairs of cyclin: CDK complexes during the cell cycle. These post-translational modifications alter Myb activity in a cell cycle dependent manner.

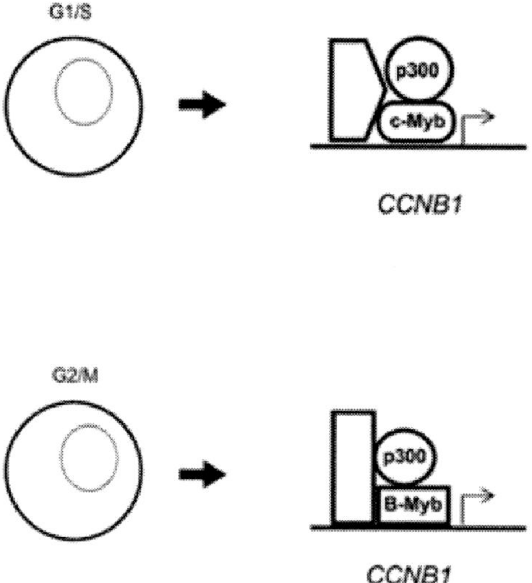

Figure 5. Myb proteins may compete or compensate for each other. B-Myb and c-Myb have some redundant functions in myeloid cells and can both bind to the CCNB1 gene. During proliferation or at different stages of the cell cycle, B-Myb and c-Myb may shuffle on and off of the same promoters in a cell cycle dependent manner.

Acknowledgments

The authors thank John P. O'Rourke and Ye Zhou for helpful comments. This work was supported by USPHS/NIH grants 5R01CA058443 and 5R01CA105257 (to SAN) and NIH fellowship 5F31HL090024 (to AMQ).

References

[1] Ness, SA. The Myb oncoprotein: regulating a regulator. *BiochimBiophysActa*, 1996, 1288, F123-39.

[2] Ness, SA. Myb protein specificity: evidence of a context-specific transcription factor code. *Blood Cells Mol Dis*, 2003, 31, 192-200.

[3] Beug, H; von Kirchbach, A; Doderlein, G; Conscience, JF; Graf, T. Chicken hematopoietic cells transformed by seven strains of defective avian leukemia viruses display three distinct phenotypes of differentiation. *Cell*, 1979, 18, 375-390.

[4] Beug, H; Hayman, MJ; Graf, T. Myeloblasts transformed by the avian acute leukemia virus E26 are hormone-dependent for growth and for the expression of a putative myb-containing protein, p135 E26. *EMBO J*, 1982, 1, 1069-1073.

[5] Radke, K; Beug, H; Kornfeld, S; Graf, T. Transformation of both erythroid and myeloid cells by E26, an avian leukemia virus that contains the myb gene. *Cell*, 1982, 31, 643-653.

[6] Klempnauer, KH; Bonifer, C; Sippel, AE. Identification and characterization of the protein encoded by the human c-myb proto-oncogene. *EMBO J*, 1986, 5, 1903-1911.

[7] Klempnauer, KH; Ramsay, G; Bishop, JM; Moscovici, MG; Moscovici, C; McGrath, JP; Levinson, AD. The product of the retroviral transforming gene v-myb is a truncated version of the protein encoded by the cellular oncogene c-myb. *Cell*, 1983, 33, 345-355.

[8] Beug, H; Leutz, A; Kahn, P; Graf, T. Ts mutants of E26 leukemia virus allow transformed myeloblasts, but not erythroblasts or fibroblasts, to differentiate at the nonpermissive temperature. *Cell*, 1984, 39, 579-588.

[9] Leutz, A; Beug, H; Graf, T. Purification and characterization of cMGF, a novel chicken myelomonocytic growth factor. *EMBO J*, 1984, 3, 3191-3197.

[10] Ness, SA; Beug, H; Graf, T. v-myb dominance over v-myc in doubly transformed chick myelomonocytic cells. *Cell*, 1987, 51, 41-50.

[11] Davies, J; Badiani, P; Weston, K. Cooperation of Myb and Myc proteins in T cell lymphomagenesis. *Oncogene*, 1999, 18, 3643-3647.

[12] Rosson, D. Effects of 5' and 3' truncations of the myb gene on the transforming ability of avian myeloblastosis virus (AMV). *Virology*, 1990, 175, 562-567.

[13] Kim, R; Trubetskoy, A; Suzuki, T; Jenkins, NA; Copeland, NG; Lenz, J. Genome-based identification of cancer genes by proviral tagging in mouse retrovirus-induced T-cell lymphomas. *J Virol*, 2003, 77, 2056-2062.

[14] Mucenski, ML; Gilbert, DJ; Taylor, BA; Jenkins, NA; Copeland, NG. Common sites of viral integration in lymphomas arising in AKXD recombinant inbred mouse strains. *Oncogene Res*, 1987, 2, 33-48.

[15] Bies, J; Koller, R; Hoffman, B; Amanullah, A; Mock, B; Wolff, L. MuLV-insertional mutagenesis of c-myb and Mml1 in a murine model for promonocytic leukemia. *Leukemia*, 1997, 11 Suppl 3, 247-250.

[16] Mukhopadhyaya, R; Wolff, L. New sites of proviral integration associated with murine promonocytic leukemias and evidence for alternate modes of c-myb activation. *J Virol*, 1992, 66, 6035-6044.

[17] Shen-Ong, GL; Wolff, L. Moloney murine leukemia virus-induced myeloid tumors in adult BALB/c mice: requirement of c-myb activation but lack of v-abl involvement. *J Virol*, 1987, 61, 3721-3725.

[18] Furuta, Y; Aizawa, S; Suda, Y; Ikawa, Y; Nakasgoshi, H; Nishina, Y; Ishii, S. Degeneration of skeletal and cardiac muscles in c-myb transgenic mice. *Transgenic Res*, 1993, 2, 199-207.

[19] Fu, SL; Lipsick, JS. Constitutive expression of full-length c-Myb transforms avian cells characteristic of both the monocytic and granulocytic lineages. *Cell Growth Differ*, 1997, 8, 35-45.

[20] Wang, DM; Dubendorff, JW; Woo, CH; Lipsick, JS. Functional analysis of carboxy-terminal deletion mutants of c-Myb. *J Virol*, 1999, 73, 5875-5886.

[21] Liu, F; Lei, W; O'Rourke, JP; Ness, SA. Oncogenic mutations cause dramatic, qualitative changes in the transcriptional activity of c-Myb. *Oncogene*, 2006, 25, 795-805.

[22] Golay, J; Capucci, A; Arsura, M; Castellano, M; Rizzo, V; Introna, M. Expression of c-myb and B-myb, but not A-myb, correlates with proliferation in human hematopoietic cells. *Blood*, 1991, 77, 149-158.

[23] Kastan, MB; Slamon, DJ; Civin, CI. Expression of protooncogene c-myb in normal human hematopoietic cells. *Blood*, 1989, 73, 1444-1451.

[24] Sakhinia, E; Faranghpour, M; Liu Yin, JA; Brady, G; Hoyland, JA; Byers, RJ. Routine expression profiling of microarray gene signatures in acute leukaemia by real-time PCR of human bone marrow. *Br J Haematol*, 2005, 130, 233-248.

[25] Mavilio, F; Sposi, NM; Petrini, M; Bottero, L; Marinucci, M; De Rossi, G; Amadori, S; Mandelli, F; Peschle, C. Expression of cellular oncogenes in primary cells from human acute leukemias. *Proc NatlAcadSci U S A*, 1986, 83, 4394-4398.

[26] Pelicci, PG; Lanfrancone, L; Brathwaite, MD; Wolman, SR; Dalla-Favera, R. Amplification of the c-myb oncogene in a case of human acute myelogenous leukemia. *Science*, 1984, 224, 1117-1121.

[27] Schraders, M; van Reijmersdal, SV; Kamping, EJ; van Krieken, JH; van Kessel, AG; Groenen, PJ; Hoogerbrugge, PM; Kuiper, RP. High-resolution genomic profiling of pediatric lymphoblastic lymphomas reveals subtle differences with pediatric acute lymphoblastic leukemias in the B-lineage. *Cancer Genet Cytogenet*, 2009, 191, 27-33.

[28] Kauraniemi, P; Hedenfalk, I; Persson, K; Duggan, DJ; Tanner, M; Johannsson, O; Olsson, H; Trent, JM; Isola, J; Borg, A. MYB oncogene amplification in hereditary BRCA1 breast cancer. *Cancer Res.*, 2000, 60, 5323-5328.

[29] Alitalo, K; Winqvist, R; Lin, CC; de la Chapelle, A; Schwab, M; Bishop, JM. Aberrant expression of an amplified c-myb oncogene in two cell lines from a colon carcinoma. *Proc NatlAcadSci U S A*, 1984, 81, 4534-4538.

[30] Ramsay, RG; Thompson, MA; Hayman, JA; Reid, G; Gonda, TJ; Whitehead, RH. Myb expression is higher in malignant human colonic carcinoma and premalignant adenomatous polyps than in normal mucosa. *Cell Growth Differ*, 1992, 3, 723-730.

[31] Guerin, M; Sheng, ZM; Andrieu, N; Riou, G. Strong association between c-myb and oestrogen-receptor expression in human breast cancer. *Oncogene*, 1990, 5, 131-135.

[32] Clappier, E; Cuccuini, W; Kalota, A; Crinquette, A; Cayucla, JM; Dik, WA; Langerak, AW; Montpellier, B; Nadel, B; Walrafen, P; Delattre, O; Aurias, A; Leblanc, T; Dombret, H; Gewirtz, AM; Baruchel, A; Sigaux, F; Soulier, J. The C-MYB locus is

involved in chromosomal translocation and genomic duplications in human T-cell acute leukemia (T-ALL), the translocation defining a new T-ALL subtype in very young children. *Blood*, 2007, 110, 1251-1261.

[33] Lahortiga, I; De Keersmaecker, K; Van Vlierberghe, P; Graux, C; Cauwelier, B; Lambert, F; Mentens, N; Beverloo, HB; Pieters, R; Speleman, F; Odero, MD; Bauters, M; Froyen, G; Marynen, P; Vandenberghe, P; Wlodarska, I; Meijerink, JP; Cools, J. Duplication of the MYB oncogene in T cell acute lymphoblastic leukemia. *Nat Genet*, 2007, 39, 593-595.

[34] Kagawa, N; Maruno, M; Suzuki, T; Hashiba, T; Hashimoto, N; Izumoto, S; Yoshimine, T. Detection of genetic and chromosomal aberrations in medulloblastomas and primitive neuroectodermal tumors with DNA microarrays. *Brain Tumor Pathol*, 2006, 23, 41-47.

[35] Griffin, CA; Baylin, SB. Expression of the c-myb oncogene in human small cell lung carcinoma. *Cancer Res.*, 1985, 45, 272-275.

[36] Liu, CJ; Lin, SC; Chen, YJ; Chang, KM; Chang, KW. Array-comparative genomic hybridization to detect genomewide changes in microdissected primary and metastatic oral squamous cell carcinomas. *Mol Carcinog*, 2006, 45, 721-731.

[37] Persson, M; Andren, Y; Mark, J; Horlings, HM; Persson, F; Stenman, G. Recurrent fusion of MYB and NFIB transcription factor genes in carcinomas of the breast and head and neck. *Proc NatlAcadSci U S A*, 2009, 106, 18740-18744.

[38] Mucenski, ML; McLain, K; Kier, AB; Swerdlow, SH; Schreiner, CM; Miller, TA; Pietryga, DW; Scott, WJJ; Potter, SS. A functional c-myb gene is required for normal murine fetal hepatic hematopoiesis. *Cell*, 1991, 65, 677-689.

[39] Emambokus, N; Vegiopoulos, A; Harman, B; Jenkinson, E; Anderson, G; Frampton, J. Progression through key stages of haemopoiesis is dependent on distinct threshold levels of c-Myb. *EMBO J*, 2003, 22, 4478-4488.

[40] Thompson, CB; Challoner, PB; Neiman, PE; Groudine, M. Expression of the c-myb proto-oncogene during cellular proliferation. *Nature*, 1986, 319, 374-380.

[41] Ramsay, RG; Ikeda, K; Rifkind, RA; Marks, PA. Changes in gene expression associated with induced differentiation of erythroleukemia: protooncogenes, globin genes, and cell division. *Proc NatlAcadSci U S A*, 1986, 83, 6849-6853.

[42] Marks, PA; Ramsay, R; Sheffery, M; Rifkind, RA. Changes in gene expression during hexamethylenebisacetamide induced erythroleukemia differentiation. *ProgClinBiol Res.*, 1987, 251, 253-268.

[43] Clarke, MF; Kukowska-Latallo, JF; Westin, E; Smith, M; Prochownik, EV. Constitutive expression of a c-myb cDNA blocks Friend murine erythroleukemia cell differentiation. *Mol Cell Biol.*, 1988, 8, 884-892.

[44] Baserga, R; Reiss, K; Alder, H; Pietrzkowski, Z; Surmacz, E. Inhibition of cell cycle progression by antisense oligodeoxynucleotides. *Ann N Y AcadSci.*, 1992, 660, 64-69.

[45] Citro, G; Perrotti, D; Cucco, C; D'Agnano, I; Sacchi, A; Zupi, G; Calabretta, B. Inhibition of leukemia cell proliferation by receptor-mediated uptake of c-myb antisense oligodeoxynucleotides. *Proc NatlAcadSci U S A*, 1992, 89, 7031-7035.

[46] Gewirtz, AM; Calabretta, B. A c-myb antisense oligodeoxynucleotide inhibits normal human hematopoiesis in vitro. *Science*, 1988, 242, 1303-1306.

[47] Kamano, H; Ohnishi, H; Tanaka, T; Ikeda, K; Okabe, A; Irino, S. Effects of the antisense v-myb' expression on K562 human leukemia cell proliferation and differentiation. *Leuk Res.*, 1990, 14, 831-839.

[48] Ratajczak, MZ; Kant, JA; Luger, SM; Hijiya, N; Zhang, J; Zon, G; Gewirtz, AM. In vivo treatment of human leukemia in a scid mouse model with c-myb antisense oligodeoxynucleotides. *Proc NatlAcadSci U S A*, 1992, 89, 11823-11827.

[49] Muller, C; Yang, R; Idos, G; Tidow, N; Diederichs, S; Koch, OM; Verbeek, W; Bender, TP; Koeffler, HP. c-myb transactivates the human cyclin A1 promoter and induces cyclin A1 gene expression. *Blood*, 1999, 94, 4255-4262.

[50] Nakata, Y; Shetzline, S; Sakashita, C; Kalota, A; Rallapalli, R; Rudnick, SI; Zhang, Y; Emerson, SG; Gewirtz, AM. c-Myb contributes to G2/M cell cycle transition in human hematopoietic cells by direct regulation of cyclin B1 expression. *Mol Cell Biol.*, 2007, 27, 2048-2058.

[51] Ganter, B; Fu, S; Lipsick, JS. D-type cyclins repress transcriptional activation by the v-Myb but not the c-Myb DNA-binding domain. *EMBO J*, 1998, 17, 255-268.

[52] Lei, W; Liu, F; Ness, SA. Positive and negative regulation of c-Myb by cyclin D1, cyclin-dependent kinases, and p27 Kip1. *Blood*, 2005, 105, 3855-3861.

[53] Malaterre, J; Carpinelli, M; Ernst, M; Alexander, W; Cooke, M; Sutton, S; Dworkin, S; Heath, JK; Frampton, J; McArthur, G; Clevers, H; Hilton, D; Mantamadiotis, T; Ramsay, RG. c-Myb is required for progenitor cell homeostasis in colonic crypts. *Proc NatlAcadSci U S A*, 2007, 104, 3829-3834.

[54] Drabsch, Y; Hugo, H; Zhang, R; Dowhan, DH; Miao, YR; Gewirtz, AM; Barry, SC; Ramsay, RG; Gonda, TJ. Mechanism of and requirement for estrogen-regulated MYB expression in estrogen-receptor-positive breast cancer cells. *Proc NatlAcadSci U S A*, 2007, 104, 13762-13767.

[55] Ness, SA; Marknell, A; Graf, T. The v-myb oncogene product binds to and activates the promyelocyte-specific mim-1 gene. *Cell*, 1989, 59, 1115-1125.

[56] Ness, SA; Kowenz-Leutz, E; Casini, T; Graf, T; Leutz, A. Myb and NF-M: combinatorial activators of myeloid genes in heterologous cell types. *Genes Dev.*, 1993, 7, 749-759.

[57] Nuchprayoon, I; Simkevich, CP; Luo, M; Friedman, AD; Rosmarin, AG. GABP cooperates with c-Myb and C/EBP to activate the neutrophil elastase promoter. *Blood*, 1997, 89, 4546-4554.

[58] Oelgeschlager, M; Nuchprayoon, I; Luscher, B; Friedman, AD. C/EBP, c-Myb, and PU.1 cooperate to regulate the neutrophil elastase promoter. *Mol Cell Biol.*, 1996, 16, 4717-4725.

[59] Britos-Bray, M; Friedman, AD. Core binding factor cannot synergistically activate the myeloperoxidase proximal enhancer in immature myeloid cells without c-Myb. *Mol Cell Biol.*, 1997, 17, 5127-5135.

[60] Melotti, P; Ku, DH; Calabretta, B. Regulation of the expression of the hematopoietic stem cell antigen CD34: role of c-myb. *J Exp Med*, 1994, 179, 1023-1028.

[61] Chu, TY; Besmer, P. Characterization of the promoter of the proto-oncogene c-kit. *Proc NatlSciCouncRepub China B*, 1995, 19, 8-18.

[62] Yamamoto, K; Tojo, A; Aoki, N; Shibuya, M. Characterization of the promoter region of the human c-kit proto-oncogene. *Jpn J Cancer Res.*, 1993, 84, 1136-1144.

[63] Todokoro, K; Watson, RJ; Higo, H; Amanuma, H; Kuramochi, S; Yanagisawa, H; Ikawa, Y. Down-regulation of c-myb gene expression is a prerequisite for erythropoietin-induced erythroid differentiation. *Proc NatlAcadSci U S A*, 1988, 85, 8900-8904.

[64] Wang, QF; Lauring, J; Schlissel, MS. c-Myb binds to a sequence in the proximal region of the RAG-2 promoter and is essential for promoter activity in T-lineage cells. *Mol Cell Biol.*, 2000, 20, 9203-9211.

[65] Kishi, H; Jin, ZX; Wei, XC; Nagata, T; Matsuda, T; Saito, S; Muraguchi, A. Cooperative binding of c-Myb and Pax-5 activates the RAG-2 promoter in immature B cells. *Blood*, 2002, 99, 576-583.

[66] Dash, AB; Orrico, FC; Ness, SA. The EVES motif mediates both intermolecular and intramolecular regulation of c-Myb. *Genes Dev.*, 1996, 10, 1858-1869.

[67] Leverson, JD; Ness, SA. Point mutations in v-Myb disrupt a cyclophilin-catalyzed negative regulatory mechanism. *Mol Cell*, 1998, 1, 203-211.

[68] Winn, LM; Lei, W; Ness, SA. Pim-1 phosphorylates the DNA binding domain of c-Myb. *Cell Cycle*, 2003, 2, 258-262.

[69] Introna, M; Golay, J; Frampton, J; Nakano, T; Ness, SA; Graf, T. Mutations in v-myb alter the differentiation of myelomonocytic cells transformed by the oncogene. *Cell*, 1990, 63, 1289-1297.

[70] Lei, W; Rushton, JJ; Davis, LM; Liu, F; Ness, SA. Positive and negative determinants of target gene specificity in myb transcription factors. *J BiolChem.*, 2004, 279, 29519-29527.

[71] Kalkbrenner, F; Guehmann, S; Moelling, K. Transcriptional activation by human c-myb and v-myb genes. *Oncogene*, 1990, 5, 657-661.

[72] Bortner, DM; Ostrowski, MC. Analysis of the v-myb structural components important for transactivation of gene expression. *Nucleic Acids Res*, 1991, 19, 1533-1539.

[73] Dubendorff, JW; Whittaker, LJ; Eltman, JT; Lipsick, JS. Carboxy-terminal elements of c-Myb negatively regulate transcriptional activation in cis and in trans. *Genes Dev*, 1992, 6, 2524-2535.

[74] Boyer, LA; Langer, MR; Crowley, KA; Tan, S; Denu, JM; Peterson, CL. Essential role for the SANT domain in the functioning of multiple chromatin remodeling enzymes. *Mol Cell*, 2002, 10, 935-942.

[75] Boyer, LA; Latek, RR; Peterson, CL. The SANT domain: a unique histone-tail-binding module? *Nat Rev Mol Cell Biol*, 2004, 5, 158-163.

[76] Mo, X; Kowenz-Leutz, E; Laumonnier, Y; Xu, H; Leutz, A. Histone H3 tail positioning and acetylation by the c-Myb but not the v-Myb DNA-binding SANT domain. *Genes Dev*, 2005, 19, 2447-2457.

[77] Ko, ER; Ko, D; Chen, C; Lipsick, JS. A conserved acidic patch in the Myb domain is required for activation of an endogenous target gene and for chromatin binding. *Mol Cancer*, 2008, 7, 77.

[78] Jin, S; Zhao, H; Yi, Y; Nakata, Y; Kalota, A; Gewirtz, AM. c-Myb binds MLL through menin in human leukemia cells and is an important driver of MLL-associated leukemogenesis. *J Clin Invest*, 2010, 120, 593-606.

[79] Biedenkapp, H; Borgmeyer, U; Sippel, AE; Klempnauer, KH. Viral myb oncogene encodes a sequence-specific DNA-binding activity. *Nature*, 1988, 335, 835-837.

[80] Rushton, JJ; Davis, LM; Lei, W; Mo, X; Leutz, A; Ness, SA. Distinct changes in gene expression induced by A-Myb, B-Myb and c-Myb proteins. *Oncogene*, 2003, 22, 308-313.

[81] Inoue, K; Mallakin, A; Frazier, DP. Dmp1 and tumor suppression. *Oncogene*, 2007, 26, 4329-4335.

[82] Mallakin, A; Sugiyama, T; Kai, F; Taneja, P; Kendig, RD; Frazier, DP; Maglic, D; Matise, LA; Willingham, MC; Inoue, K. The Arf-inducing transcription factor Dmp1 encodes a transcriptional activator of amphiregulin, thrombospondin-1, JunB and Egr1. *Int J Cancer*, 2010, 126, 1403-1416.

[83] Bilaud, T; Koering, CE; Binet-Brasselet, E; Ancelin, K; Pollice, A; Gasser, SM; Gilson, E. The telobox, a Myb-related telomeric DNA binding motif found in proteins from yeast, plants and human. *Nucleic Acids Res.*, 1996, 24, 1294-1303.

[84] Broccoli, D; Smogorzewska, A; Chong, L; de Lange, T. Human telomeres contain two distinct Myb-related proteins, TRF1 and TRF2. *Nat Genet*, 1997, 17, 231-235.

[85] Tahmaseb, K; Turchi, JJ. Intrinsic hTRF1 fluorescence quenching reveals details of telomere DNA binding activity: impact of DNA length, structure and position of telomeric repeats. *Arch BiochemBiophys*, 2010, 493, 207-212.

[86] Lee, TH; Tun-Kyi, A; Shi, R; Lim, J; Soohoo, C; Finn, G; Balastik, M; Pastorino, L; Wulf, G; Zhou, XZ; Lu, KP. Essential role of Pin1 in the regulation of TRF1 stability and telomere maintenance. *Nat Cell Biol.*, 2009, 11, 97-105.

[87] Pani, E; Menigatti, M; Schubert, S; Hess, D; Gerrits, B; Klempnauer, KH; Ferrari, S. Pin1 interacts with c-Myb in a phosphorylation-dependent manner and regulates its transactivation activity. *BiochimBiophysActa*, 2008, 1783, 1121-1128.

[88] Bousset, K; Oelgeschlager, MH; Henriksson, M; Schreek, S; Burkhardt, H; Litchfield, DW; Luscher-Firzlaff, JM; Luscher, B. Regulation of transcription factors c-Myc, Max, and c-Myb by casein kinase II. *Cell Mol Biol Res.*, 1994, 40, 501-511.

[89] Luscher, B; Christenson, E; Litchfield, DW; Krebs, EG; Eisenman, RN. Myb DNA binding inhibited by phosphorylation at a site deleted during oncogenic activation. *Nature*, 1990, 344, 517-522.

[90] Andersson, K; B; Kowenz-Leutz, E; Brendeford, EM; Tygsett, AH; Leutz, A; Gabrielsen, OS. Phosphorylation-dependent down-regulation of c-Myb DNA binding is abrogated by a point mutation in the v-myb oncogene. *J BiolChem.*, 2003, 278, 3816-3824.

[91] Aziz, N; Miglarese, MR; Hendrickson, RC; Shabanowitz, J; Sturgill, TW; Hunt, DF; Bender, TP. Modulation of c-Myb-induced transcription activation by a phosphorylation site near the negative regulatory domain. *Proc NatlAcadSci U S A*, 1995, 92, 6429-6433.

[92] Aziz, N; Wu, J; Dubendorff, JW; Lipsick, JS; Sturgill, TW; Bender, TP. c-Myb and v-Myb are differentially phosphorylated by p42mapk in vitro. *Oncogene*, 1993, 8, 2259-2265.

[93] Bies, J; Feikova, S; Bottaro, DP; Wolff, L. Hyperphosphorylation and increased proteolytic breakdown of c-Myb induced by the inhibition of Ser/Thr protein phosphatases. *Oncogene*, 2000, 19, 2846-2854.

[94] Bies, J; Feikova, S; Markus, J; Wolff, L. Phosphorylation-dependent conformation and proteolytic stability of c-Myb. *Blood Cells Mol Dis.*, 2001, 27, 422-428.

[95] Kanei-Ishii, C; Ninomiya-Tsuji, J; Tanikawa, J; Nomura, T; Ishitani, T; Kishida, S; Kokura, K; Kurahashi, T; Ichikawa-Iwata, E; Kim, Y; Matsumoto, K; Ishii, S. Wnt-1 signal induces phosphorylation and degradation of c-Myb protein via TAK1, HIPK2, and NLK. *Genes Dev*, 2004, 18, 816-829.

[96] Kitagawa, K; Hiramatsu, Y; Uchida, C; Isobe, T; Hattori, T; Oda, T; Shibata, K; Nakamura, S; Kikuchi, A; Kitagawa, M. Fbw7 promotes ubiquitin-dependent degradation of c-Myb: involvement of GSK3-mediated phosphorylation of Thr-572 in mouse c-Myb. *Oncogene*, 2009, 28, 2393-2405.

[97] Matre, V; Nordgard, O; Alm-Kristiansen, AH; Ledsaak, M; Gabrielsen, OS. HIPK1 interacts with c-Myb and modulates its activity through phosphorylation. *BiochemBiophys Res Commun*, 2009, 388, 150-154.

[98] Miglarese, MR; Richardson, AF; Aziz, N; Bender, TP. Differential regulation of c-Myb-induced transcription activation by a phosphorylation site in the negative regulatory domain. *J BiolChem.*, 1996, 271, 22697-22705.

[99] Pani, E; Ferrari, S. p38MAPK delta controls c-Myb degradation in response to stress. *Blood Cells Mol Dis.*, 2008, 40, 388-394.

[100] Vorbrueggen, G; Lovric, J; Moelling, K. Functional analysis of phosphorylation at serine 532 of human c-Myb by MAP kinase. *BiolChem.*, 1996, 377, 721-730.

[101] Bies, J; Markus, J; Wolff, L. Covalent attachment of the SUMO-1 protein to the negative regulatory domain of the c-Myb transcription factor modifies its stability and transactivation capacity. *J BiolChem.*, 2002, 277, 8999-9009.

[102] Dahle, O; Andersen, TO; Nordgard, O; Matre, V; Del Sal, G; Gabrielsen, OS. Transactivation properties of c-Myb are critically dependent on two SUMO-1 acceptor sites that are conjugated in a PIASy enhanced manner. *Eur J Biochem*, 2003, 270, 1338-1348.

[103] Sramko, M; Markus, J; Kabat, J; Wolff, L; Bies, J. Stress-induced inactivation of the c-Myb transcription factor through conjugation of SUMO-2/3 proteins. *J BiolChem.*, 2006, 281, 40065-40075.

[104] Corradini, F; Cesi, V; Bartella, V; Pani, E; Bussolari, R; Candini, O; Calabretta, B. Enhanced proliferative potential of hematopoietic cells expressing degradation-resistant c-Myb mutants. *J BiolChem.*, 2005, 280, 30254-30262.

[105] Sandburg, SL; Sutton, SE; Pletcher, MT; Wiltshire, T; Tarantino, LM; Hogenesch, JB; Cooke, MP. c-Myb and p300 regulate hematopoietic stem cell proliferation and differentiation. *Dev Cell*, 2005, 8, 153-166.

[106] Tomita, A; Towatari, M; Tsuzuki, S; Hayakawa, F; Kosugi, H; Tamai, K; Miyazaki, T; Kinoshita, T; Saito, H. c-Myb acetylation at the carboxyl-terminal conserved domain by transcriptional co-activator p300. *Oncogene*, 2000, 19, 444-451.

[107] Dai, P; Akimaru, H; Tanaka, Y; Hou, DX; Yasukawa, T; Kanei-Ishii, C; Takahashi, T; Ishii, S. CBP as a transcriptional coactivator of c-Myb. *Genes Dev*, 1996, 10, 528-540.

[108] Sano, Y; Ishii, S. Increased affinity of c-Myb for CREB-binding protein (CBP) after CBP-induced acetylation. *J BiolChem.*, 2001, 276, 3674-3682.

[109] Muller-Tidow, C; Wang, W; Idos, GE; Diederichs, S; Yang, R; Readhead, C; Berdel, WE; Serve, H; Saville, M; Watson, R; Koeffler, HP. Cyclin A1 directly interacts with B-myb and cyclin A1/cdk2 phosphorylate B-myb at functionally important serine and threonine residues: tissue-specific regulation of B-myb function. *Blood*, 2001, 97, 2091-2097.

[110] Feikova, S; Wolff, L; Bies, J. Constitutive ubiquitination and degradation of c-myb by the 26S proteasome during proliferation and differentiation of myeloid cells. *Neoplasma*, 2000, 47, 212-218.

[111] Dudek, H; Reddy, EP. Identification of two translational products for c-myb. *Oncogene*, 1989, 4, 1061-1066.

[112] O'Rourke, JP; Ness, SA. Alternative RNA splicing produces multiple forms of c-Myb with unique transcriptional activities. *Mol Cell Biol.*, 2008, 28, 2091-2101.

[113] Shen-Ong, GL. Alternative internal splicing in c-myb RNAs occurs commonly in normal and tumor cells. *EMBO J*, 1987, 6, 4035-4039.

[114] Westin, EH; Gorse, KM; Clarke, MF. Alternative splicing of the human c-myb gene. *Oncogene*, 1990, 5, 1117-1124.

[115] Schuur, ER; Dasgupta, P; Reddy, EP; Rabinovich, JM; Baluda, MA. Alternative splicing of the chicken c-myb exon 9A. *Oncogene*, 1993, 8, 1839-1847.

[116] Schuur, ER; Rabinovich, JM; Baluda, MA. Distribution of alternatively spliced chicken c-myb exon 9A among hematopoietic tissues. *Oncogene*, 1994, 9, 3363-3365.

[117] Karni, R; de Stanchina, E; Lowe, SW; Sinha, R; Mu, D; Krainer, AR. The gene encoding the splicing factor SF2/ASF is a proto-oncogene. *Nat Struct Mol Biol.*, 2007, 14, 185-193.

[118] Karni, R; Hippo, Y; Lowe, SW; Krainer, AR. The splicing-factor oncoprotein SF2/ASF activates mTORC1. *Proc NatlAcadSci U S A*, 2008, 105, 15323-15327.

[119] Lu, YD; Gan, QH; Chi, XY; Qin, S. Roles of microRNA in plant defense and virus offense interaction. *Plant Cell Rep.*, 2008, 27, 1571-1579.

[120] Bartel, DP. MicroRNAs: genomics, biogenesis, mechanism, and function. *Cell*, 2004, 116, 281-297.

[121] O'Day, E; Lal, A. MicroRNAs and their target gene networks in breast cancer. *Breast Cancer Res.*, 2010, 12, 201.

[122] Jakymiw, A; Patel, RS; Deming, N; Bhattacharyya, I; Shah, P; Lamont, RJ; Stewart, CM; Cohen, DM; Chan, EK. Overexpression of dicer as a result of reduced let-7 MicroRNA levels contributes to increased cell proliferation of oral cancer cells. *Genes Chromosomes Cancer*, 2010.

[123] Venugopal, SK; Jiang, J; Kim, TH; Li, Y; Wang, SS; Torok, NJ; Wu, J; Zern, MA. Liver fibrosis causes down-regulation of miRNA-150 and miRNA-194 in hepatic stellate cells and their over-expression causes decreased stellate cell activation. *Am J PhysiolGastrointest Liver Physiol*, 2009.

[124] Israel, A; Sharan, R; Ruppin, E; Galun, E. Increased microRNA activity in human cancers. *PLoS One*, 2009, 4, e6045.

[125] Monticelli, S; Ansel, KM; Xiao, C; Socci, ND; Krichevsky, AM; Thai, TH; Rajewsky, N; Marks, DS; Sander, C; Rajewsky, K; Rao, A; Kosik, KS. MicroRNA profiling of the murine hematopoietic system. *Genome Biol.*, 2005, 6, R71.

[126] Xiao, C; Calado, DP; Galler, G; Thai, TH; Patterson, HC; Wang, J; Rajewsky, N; Bender, TP; Rajewsky, K. MiR-150 controls B cell differentiation by targeting the transcription factor c-Myb. *Cell*, 2007, 131, 146-159.

[127] Lin, YC; Kuo, MW; Yu, J; Kuo, HH; Lin, RJ; Lo, WL; Yu, AL. c-Myb is an evolutionary conserved miR-150 target and miR-150/c-Myb interaction is important for embryonic development. *Mol BiolEvol*, 2008, 25, 2189-2198.

[128] Lu, J; Guo, S; Ebert, BL; Zhang, H; Peng, X; Bosco, J; Pretz, J; Schlanger, R; Wang, JY; Mak, RH; Dombkowski, DM; Preffer, FI; Scadden, DT; Golub, TR. MicroRNA-mediated control of cell fate in megakaryocyte-erythrocyte progenitors. *Dev Cell*, 2008, 14, 843-853.

[129] Garcia, P; Clarke, M; Vegiopoulos, A; Berlanga, O; Camelo, A; Lorvellec, M; Frampton, J. Reduced c-Myb activity compromises HSCs and leads to a myeloproliferation with a novel stem cell basis. *EMBO J*, 2009, 28, 1492-1504.

[130] Chung, EY; Dews, M; Cozma, D; Yu, D; Wentzel, EA; Chang, TC; Schelter, JM; Cleary, MA; Mendell, JT; Thomas-Tikhonenko, A. c-Myb oncoprotein is an essential target of the dleu2 tumor suppressor microRNA cluster. *Cancer BiolTher.*, 2008, 7, 1758-1764.

[131] Zhao, H; Kalota, A; Jin, S; Gewirtz, AM. The c-myb proto-oncogene and microRNA-15a comprise an active autoregulatory feedback loop in human hematopoietic cells. *Blood*, 2009, 113, 505-516.

[132] Navarro, F; Gutman, D; Meire, E; Caceres, M; Rigoutsos, I; Bentwich, Z; Lieberman, J. miR-34a contributes to megakaryocytic differentiation of K562 cells independently of p53. *Blood*, 2009, 114, 2181-2192.

[133] Lerner, M; Harada, M; Loven, J; Castro, J; Davis, Z; Oscier, D; Henriksson, M; Sangfelt, O; Grander, D; Corcoran, MM. DLEU2, frequently deleted in malignancy, functions as a critical host gene of the cell cycle inhibitory microRNAs miR-15a and miR-16-1. *Exp Cell Res.*, 2009, 315, 2941-2952.

[134] Lu, Y; Ryan, SL; Elliott, DJ; Bignell, GR; Futreal, PA; Ellison, DW; Bailey, S; Clifford, SC. Amplification and overexpression of Hsa-miR-30b, Hsa-miR-30d and KHDRBS3 at 8q24.22-q24.23 in medulloblastoma. *PLoS One*, 2009, 4, e6159.

[135] Garzon, R; Calin, GA; Croce, CM. MicroRNAs in Cancer. *Annu Rev Med*, 2009, 60, 167-179.

[136] Lee, J; Li, Z; Brower-Sinning, R; John, B. Regulatory circuit of human microRNA biogenesis. *PLoSComputBiol*, 2007, 3, e67.

[137] Bender, TP; Thompson, CB; Kuehl, WM. Differential expression of c-myb mRNA in murine B lymphomas by a block to transcription elongation. *Science*, 1987, 237, 1473-1476.

[138] Watson, RJ. A transcriptional arrest mechanism involved in controlling constitutive levels of mouse c-myb mRNA. *Oncogene*, 1988, 2, 267-272.

[139] Reddy, CD; Reddy, EP. Differential binding of nuclear factors to the intron 1 sequences containing the transcriptional pause site correlates with c-myb expression. *Proc NatlAcadSci U S A*, 1989, 86, 7326-7330.

[140] Hugo, H; Cures, A; Suraweera, N; Drabsch, Y; Purcell, D; Mantamadiotis, T; Phillips, W; Dobrovic, A; Zupi, G; Gonda, TJ; Iacopetta, B; Ramsay, RG. Mutations in the MYB intron I regulatory sequence increase transcription in colon cancers. *Genes Chromosomes Cancer*, 2006, 45, 1143-1154.

[141] Bender, TP; Kremer, CS; Kraus, M; Buch, T; Rajewsky, K. Critical functions for c-Myb at three checkpoints during thymocyte development. *Nat Immunol*, 2004, 5, 721-729.

[142] Ku, DH; Wen, SC; Engelhard, A; Nicolaides, NC; Lipson, KE; Marino, TA; Calabretta, B. c-myb transactivates cdc2 expression via Myb binding sites in the 5'-flanking region of the human cdc2 gene. *J BiolChem.*, 1993, 268, 2255-2259.

[143] Salomoni, P; Perrotti, D; Martinez, R; Franceschi, C; Calabretta, B. Resistance to apoptosis in CTLL-2 cells constitutively expressing c-Myb is associated with induction of BCL-2 expression and Myb-dependent regulation of bcl-2 promoter activity. *Proc NatlAcadSci U S A*, 1997, 94, 3296-3301.

[144] Thompson, MA; Rosenthal, MA; Ellis, SL; Friend, AJ; Zorbas, MI; Whitehead, RH; Ramsay, RG. c-Myb down-regulation is associated with human colon cell differentiation, apoptosis, and decreased Bcl-2 expression. *Cancer Res.*, 1998, 58, 5168-5175.

[145] Yuan, J; Crittenden, RB; Bender, TP. c-Myb promotes the survival of CD4+CD8+ double-positive thymocytes through upregulation of Bcl-xL. *J Immunol*, 2010, 184, 2793-2804.

[146] Fahl, SP; Crittenden, RB; Allman, D; Bender, TP. c-Myb is required for pro-B cell differentiation. *J Immunol*, 2009, 183, 5582-5592.

[147] Alm-Kristiansen, AH; Saether, T; Matre, V; Gilfillan, S; Dahle, O; Gabrielsen, OS. FLASH acts as a co-activator of the transcription factor c-Myb and localizes to active RNA polymerase II foci. *Oncogene*, 2008, 27, 4644-4656.

[148] Gewirtz, AM; Anfossi, G; Venturelli, D; Valpreda, S; Sims, R; Calabretta, B. G1/S transition in normal human T-lymphocytes requires the nuclear protein encoded by c-myb. *Science*, 1989, 245, 180-183.

[149] Torelli, G; Selleri, L; Donelli, A; Ferrari, S; Emilia, G; Venturelli, D; Moretti, L; Torelli, U. Activation of c-myb expression by phytohemagglutinin stimulation in normal human T lymphocytes. *Mol Cell Biol.*, 1985, 5, 2874-2877.

[150] Lipsick, JS; Boyle, WJ. c-myb protein expression is a late event during T-lymphocyte activation. *Mol Cell Biol.*, 1987, 7, 3358-3360.

[151] Campanero, MR; Armstrong, M; Flemington, E. Distinct cellular factors regulate the c-myb promoter through its E2F element. *Mol Cell Biol.*, 1999, 19, 8442-8450.

[152] Lam, EW; Bennett, JD; Watson, RJ. Cell-cycle regulation of human B-myb transcription. *Gene*, 1995, 160, 277-281.

[153] Marhamati, DJ; Bellas, RE; Arsura, M; Kypreos, KE; Sonenshein, GE. A-myb is expressed in bovine vascular smooth muscle cells during the late G1-to-S phase transition and cooperates with c-myc to mediate progression to S phase. *Mol Cell Biol.*, 1997, 17, 2448-2457.

[154] Davidson, CJ; Tirouvanziam, R; Herzenberg, LA; Lipsick, JS. Functional evolution of the vertebrate Myb gene family: B-Myb, but neither A-Myb nor c-Myb, complements Drosophila Myb in hemocytes. *Genetics*, 2005, 169, 215-229.

[155] Fitzpatrick, CA; Sharkov, NV; Ramsay, G; Katzen, AL. Drosophila myb exerts opposing effects on S phase, promoting proliferation and suppressing endoreduplication. *Development*, 2002, 129, 4497-4507.

[156] Fung, SM; Ramsay, G; Katzen, AL. Mutations in Drosophila myb lead to centrosome amplification and genomic instability. *Development*, 2002, 129, 347-359.

[157] Katzen, AL; Jackson, J; Harmon, BP; Fung, SM; Ramsay, G; Bishop, JM. Drosophila myb is required for the G2/M transition and maintenance of diploidy. *Genes Dev.*, 1998, 12, 831-843.

[158] Pilkinton, M; Sandoval, R; Colamonici, OR. Mammalian Mip/LIN-9 interacts with either the p107, p130/E2F4 repressor complex or B-Myb in a cell cycle-phase-

dependent context distinct from the Drosophila dREAM complex. *Oncogene*, 2007, 26, 7535-7543.

[159] Schmit, F; Korenjak, M; Mannefeld, M; Schmitt, K; Franke, C; von Eyss, B; Gagrica, S; Hanel, F; Brehm, A; Gaubatz, S. LINC, a human complex that is related to pRB-containing complexes in invertebrates regulates the expression of G2/M genes. *Cell Cycle*, 2007, 6, 1903-1913.

[160] Yamauchi, T; Ishidao, T; Nomura, T; Shinagawa, T; Tanaka, Y; Yonemura, S; Ishii, S. A B-Myb complex containing clathrin and filamin is required for mitotic spindle function. *EMBO J*, 2008, 27, 1852-1862.

[161] Mannefeld, M; Klassen, E; Gaubatz, S. B-MYB is required for recovery from the DNA damage-induced G2 checkpoint in p53 mutant cells. *Cancer Res.*, 2009, 69, 4073-4080.

[162] Knight, AS; Notaridou, M; Watson, RJ. A Lin-9 complex is recruited by B-Myb to activate transcription of G2/M genes in undifferentiated embryonal carcinoma cells. *Oncogene*, 2009, 28, 1737-1747.

[163] Tanaka, Y; Patestos, NP; Maekawa, T; Ishii, S. B-myb is required for inner cell mass formation at an early stage of development. *J BiolChem.*, 1999, 274, 28067-28070.

[164] Sleeman, JP. Xenopus A-myb is expressed during early spermatogenesis. *Oncogene*, 1993, 8, 1931-1941.

[165] Toscani, A; Mettus, RV; Coupland, R; Simpkins, H; Litvin, J; Orth, J; Hatton, KS; Reddy, EP. Arrest of spermatogenesis and defective breast development in mice lacking A-myb. *Nature*, 1997, 386, 713-717.

[166] DeRocco, SE; Iozzo, R; Ma, XP; Schwarting, R; Peterson, D; Calabretta, B. Ectopic expression of A-myb in transgenic mice causes follicular hyperplasia and enhanced B lymphocyte proliferation. *Proc NatlAcadSci U S A*, 1997, 94, 3240-3244.

[167] Thorner, AR; Hoadley, KA; Parker, JS; Winkel, S; Millikan, RC; Perou, CM. In vitro and in vivo analysis of B-Myb in basal-like breast cancer. *Oncogene*, 2009, 28, 742-751.

[168] Chellappan, SP; Hiebert, S; Mudryj, M; Horowitz, JM; Nevins, JR. The E2F transcription factor is a cellular target for the RB protein. *Cell*, 1991, 65, 1053-1061.

[169] Hiebert, SW; Chellappan, SP; Horowitz, JM; Nevins, JR. The interaction of RB with E2F coincides with an inhibition of the transcriptional activity of E2F. *Genes Dev*, 1992, 6, 177-185.

[170] Gerondakis, S; Bishop, JM. Structure of the protein encoded by the chicken proto-oncogene c-myb. *Mol Cell Biol.*, 1986, 6, 3677-3684.

[171] Gonda, TJ; Bishop, JM. Structure and transcription of the cellular homolog (c-myb) of the avian myeloblastosisvirus transforming gene (v-myb). *J Virol.*, 1983, 46, 212-220.

[172] Murray-Zmijewski, F; Slee, EA; Lu, X. A complex barcode underlies the heterogeneous response of p53 to stress. *Nat Rev Mol Cell Biol.*, 2008, 9, 702-712.

[173] Polager, S; Ginsberg, D. p53 and E2f: partners in life and death. *Nat Rev Cancer*, 2009, 9, 738-748.

[174] Ziebold, U; Klempnauer, KH. Linking Myb to the cell cycle: cyclin-dependent phosphorylation and regulation of A-Myb activity. *Oncogene*, 1997, 15, 1011-1019.

In: Oncoproteins: Types and Detection
Editor: Jeremy R. Davis
ISBN: 978-1-61761-551-1

Chapter 3

Cell Surface Oncoproteomics: Cancer Biomarker Discovery and Clinical Applications

K.L. Kaufman, S. Mactier, P. Kohnke and R.I. Christopherson
Cancer Proteomics Laboratory, School of Molecular Bioscience,
The University of Sydney, Australia

Abstract

Tumour invasion and metastasis are often responsible for a high mortality rate in cancer patients. The spread of cancer involves multiple signalling pathways mediated by cell surface proteins and generally results in the loss of cell-cell adhesion leading to cell migration and invasion of distant tissues. Many transformed cells may be eliminated by immune surveillance, but others evade detection. Cell surface profiling is of special interest in oncoproteomics as it may provide a more accurate and reliable diagnosis and prognosis and identify new therapeutic targets. Such profiles may provide disease signatures that correlate with disease subtypes, reflecting the mutated genetic program of malignant cells. Extensive immunophenotypes of live-cells captured on antibody microarrays provide a quick, reproducible scan of the cell surface for patterns of known tumour biomarkers. Discovery proteomics projects, including glycoprotein-targeted approaches, provide potential biomarkers that can be screened on large clinical cohorts using antibody microarrays. The application of these technologies is discussed in the context of leukaemia and metastatic melanoma with their potential use in the clinic.

1. Introduction

Tumour invasion and metastasis is often responsible for a high mortality rate in cancer patients. Early detection and surgical resection remains the most important treatment and minimises the risk of disease progression and death. Yet, for many cancers, early symptoms can be vague and remain subclinical until the cancer is in a more advanced stage. However, for some, like metastatic melanoma for example, there is a spectrum of responses amongst Stage III patients, with a small sub-set experiencing a low relapse rate and survival greater than 4 years, while other identical stage metastatic melanoma patients have a survival rate of less than 12 months. This not only highlights the need for more sensitive early screening tests in high-risk patients, but also the development of tools that better describe an individual's cancer characteristics so that the most effective treatment course may be selected.

To form distant metastatic colonies, tumour cells must overcome a series of complex obstacles. These primarily include resisting immune surveillance, invasion of local host tissue stroma, penetration of local lymphatic and blood vessels, extravasation and adaptation to the newly colonised microenvironment. Tumour cells therefore need to acquire molecular properties that enable the metastatic process, most of which is mediated by cell-surface proteins [1]. The identification of proteins that are preferentially expressed on the cell surface of metastatic tumour cells is therefore of fundamental importance in cancer research [2].

Oncoproteomics is the study of proteins and their interactions in a cancer cell or tissue by proteomic technologies such as 2D gel electrophoresis (2DE), liquid chromatography coupled with mass spectrometry (LC-MS) and antibody microarrays. The driving force behind oncoproteomics is the notion that different malignancies can be identified by protein signatures or patterns. Thus, a correlation between clinical parameters and defined protein expression patterns could be used for early and predictive diagnosis, including the risk for tumour relapse, disease progression and survival [3]. Such signatures may also enable tumour stratification, classifying new tumour subclasses with their own inherent risks and probable outcomes. Oncoproteomics also has the potential to revolutionize clinical practice, selecting therapeutic modalities on an individual case basis. This may include cancer diagnosis based on proteomic analysis platforms as a complement to histopathology, individualised selection of combination therapeutics that target cancer-specific protein networks, real-time assessment of therapeutic efficacy and toxicity and rational modulation of therapy based on changes in the cancer protein network associated with prognosis and drug resistance [3].

Many proteomic approaches aim to describe the metastatic potential of cancer cells, which is of particular importance considering that the majority of cancer-related deaths are associated with the metastatic spread of the disease [4]. Moreover, due to the importance of cell surface proteins in metastasis, profiling the cell surface sub-proteome of specific tumour cell types at specific differentiation or disease stages has great potential for identifying novel targets for diagnostics and therapeutics [5]. Such novel cell surface markers could be exploited for antibody-based therapy, vaccine development or other forms of immunotherapy [6]. Signaling pathways initiated by cell surface membrane proteins or receptors could also be targeted for drug-based therapy. Yet, the technological challenges are significant; the human proteome is not fully defined. No universal proteome analysis methodology has been exemplified and due to the complexity of the proteome, enrichment and fractionation methods are required to detect the significant, but often low-abundance disease-related proteins. Moreover, conventional proteomic approaches that rely on 2DE encounter difficulties analyzing membrane associated and low abundance proteins [2].

This chapter will review the involvement of cell surface proteins in immune surveillance and tumour cell invasion and metastasis (summarised in Table 1), highlighting examples from common human cancers. Conventional discovery proteomics technologies will be introduced as well as their compatibility with analysing the cell surface sub-proteome. The application of these technologies to leukaemia and metastatic melanoma is discussed with their potential uses in the clinic.

2. Cell Surface Proteins in Tumorigenesis and Metastasis

2.1. Immune Surveillance

Escaping immune surveillance is a trait shared by many common human cancers. Natural killer (NK) cells are one of the most potent modes of defence against cancers [7]. Natural killer group 2D (NKG2D) is a receptor on the surface of NK cells and T-cells that when engaged by tumour-associated ligands may promote tumour rejection by stimulating innate and adaptive lymphocyte responses [8, 9]. A variety of cancers express ligands for NKG2D, collectively named MHC class I chain-related (MIC) [10]. A study by Groh and colleagues showed MIC-positive tumour cells shed soluble MIC into the blood resulting in the marked reduction of NKG2D surface expression on tumour infiltrating and blood T-cells, in turn severely impairing tumour-antigen-specific effector T-cells [8].

Tumour cells expressing the trans-membrane protein CD200 can negatively influence the immune system through multiple pathways. Interaction of CD200 with its receptor CD200R, expressed on monocyte/macrophage lineage cells and T-cells [11], inhibits cells of the macrophage lineage [12, 13], altering cytokine profiles and inducing regulatory T-cells [14, 15], which are thought to hamper tumour-specific effector T-cell immunity [16]. *B-RAF* and *N-RAS* mutations, genetic lesions present in most melanomas, induce CD200, which represses dendritic cell function [17]. Overexpression of CD200 on clinical B-cell chronic lymphocytic leukaemia (B-CLL] cells [18] has lead to the development of anti-CD200 therapy to stimulate the immune system to eradicate CD200-expressing tumour cells [19].

2.2. Initial Steps of Invasion and Metastasis: Loss of Contact from the Primary Mass

The first step in the metastatic cascade is the detachment of tumour cells from the primary mass. Transformed and malignant cells that become detached from the epithelium display loss of adherens junctions, which in normal epithelial cells are regulated by E-cadherin-mediated cell-cell interaction and β1-integrin-mediated adhesion to the basement membrane [20, 21]. E-cadherin, like other cadherin family members, displays homophilic binding. Its adhesive functions are stabilized by β-catenin, which binds to its cytoplasmic domain, providing a link to α-catenin and the actin cytoskeleton [22]. Although E-cadherin mutations do occur, they are not commonly observed in malignant tumours [22]. Instead, E-cadherin function may be decreased or abrogated by transcriptional repression (mediated by Snail, Slug, SIP1, Twist, dEF1 and E12/E47) [23-27], followed by promoter methylation [28,

29], disruption of cytoskeletal connections by interference with its association with the catenins or by suppression of catenin [22].

Loss of E-cadherin in malignant cells may be replaced by other cadherins, most commonly N-cadherin. This 'cadherin switch' is associated with a phenotypic change observed *in vitro* known as the epithelial-to-mesenchymal transition (EMT), where epithelial cells adopt a motile fibroblast-like appearance and express mesenchymal rather than epithelial cell markers. EMT is proposed to reflect invasive and metastatic properties of transformed epithelial cells, but remains controversial as is difficult to observe *in vivo* [30]. Nevertheless expression of N-cadherin plays an important role in invasion by regulating fibroblast growth factor receptor (FGFR) function. This interaction triggers downstream signalling cascades, including phospholipase C-γ, phosphatidylinositol-3-kinase and mitogen-activated protein kinase (MAPK). The combined action of these signalling pathways promotes cell survival, migration and invasion.

2.3. Tumour Cell Invasion

For cancer cells to become metastatic and endothelial cells angiogenic, they must penetrate basal membranes (BM) and migrate through the stromal extracellular matrix (ECM) into surrounding tissues. This is initially mediated by the altered expression of cell surface integrins, release of proteases that remodel the ECM and deposition of new ECM molecules [31]. These changes activate signaling cascades that regulate gene expression, cytoskeletal organization, cell adhesion and cell survival. As a result, cancer cells become more invasive, migratory and better able to survive in different microenvironments [31]. Integrins, a diverse family of glycoproteins that form heterodimeric receptors for ECM molecules are also involved in regulating the activities of proteolytic enzymes, e.g., matrix metalloproteinases (MMPs) that degrade the BM. Increased MMP expression and activity has been linked to malignancy and tumor cell invasion. Dramatic differences in surface expression and distribution of integrins in malignant cells compared with pre-neoplastic tumours of the same type has also been well documented [32]. For instance, integrin αvβ3 is strongly expressed at the invasive front of malignant melanoma cells and angiogenic blood vessels, but weakly expressed on pre-neoplastic melanomas and quiescent blood vessels [33]. To exert their effects, integrins cooperate with receptor tyrosine kinases (RTKs), including epidermal growth factor receptor (EGFR), platelet-derived growth factor receptor (PDGFR), human epidermal growth factor receptor 2 (HER-2) and hepatocyte growth factor receptor (HGFR) or c-Met. By altering their integrin repertoire, neoplastic cells can hone part of the molecular machinery that underlies adhesion, migration, survival and growth to maximise their needs [30].

The EMT also results in changes in glycosylation of cell surface proteoglycans such as CD44, the principal cell surface receptor for hyaluronic acid (HA). The role of CD44 in cell-matrix interactions suggests that altered CD44 expression on neoplastic cells and its interaction with HA may relate to tumour cell proliferation [34], migration [34-37], invasion [35, 38] and formation of metastatic tumour emboli [39-43]. High levels of CD44 are associated with increased metastatic risk and poor prognosis for primary tumours of malignant melanoma [44, 45] and lymphoma [46-48]. Although many human malignancies express high levels of CD44 [49], many including gastrointestinal cancer, bladder cancer, cervical cancer, melanoma, breast cancer and non-Hodgkin's lymphoma express different patterns of several CD44 isoforms [35, 45]. The magnitude of CD44 variant synthesis at the protein level may correlate with lymph node metastasis perhaps through regulating HA binding [50].

2.4. Cell Surface Markers for Metastasis

Metastatic tumours spread to different organs resulting in the death of cancer patients [4]. Despite a tremendous wealth of information on tumour cell biology, many key questions are yet to be answered. For instance, is a metastatic proclivity established during cell transformation, or is this characteristic a result of compounding mutations and selection? The more recently emerging viewpoint, based on gene expression profile analysis of diverse primary and metastatic tumours, is that metastatic cells may constitute part of the early makeup of a malignant tumour [30]. In other words, the migration and invasion of cancer cells is not entirely random. However, why some metastatic tumours have a more restricted range of target tissues than others, for example, prostate cancer metastasis is largely confined to bone [51] and uveal melanoma is almost exclusively confined to the liver [52], is currently unknown [53]. Recent studies indicate important roles of certain cell-surface markers as indicators and/or predictors of distant metastasis, e.g. HER-2-expressing breast cancer cells are more likely to metastasise to the brain [54].

2.5. Adaptation to a Newly Colonised Microenvironment

The infiltration of distant organs by circulating cancer cells involves specialised activities that enable extravasation and survival in the newly invaded parenchyma. As the structure and composition of capillary walls and the sub-adjacent parenchyma vary in different organs, the activities required for metastatic infiltration, survival and colonisation may differ depending on the target organ [53]. For a secondary tumour to grow, circulating cancer cells must be 'captured' by a particular tissue through interactions between surface molecules. Unique endothelial cell surface molecules can act as targets for compatible circulating cancer cells [55], e.g. adhesion molecule, metadherin appears to mediate breast cancer cell metastasis to the lungs by selectively binding to the pulmonary vasculature [56] and C-X-C chemokine receptor 4 (CXCR4) is a marker and mediator of bone metastasis in breast cancer [57]. The structural features of the target organ's native capillaries can also affect cancer cell infiltration. For instance, liver capillaries are fenestrated and may be readily traversed by cancer cells compared with other organs. However, for brain metastasis, crossing the tightly regulated blood-brain barrier by cancer cells presents more of an obstacle, and could require highly specialized functions, yet to be characterised.

Irrespective of the distant organ location, metastasized cancer cells will encounter different selective pressures from those at the primary tumour site and survival of these disseminated, extravasated cancer cells relies on their ability to establish productive interactions with the new host microenvironment. The specific molecular mediators of colonization of other organ microenvironments remain unknown [53]. However, this is likely to involve an active crosstalk via complex signaling pathways between cancer cells, immune cells and the various cell types of the host organ, some of which are likely mediated by cell surface proteins.

Table 1. Summary of cell surface proteins important in cancer

Cell surface molecules	Type of interaction	Role in cancer	Ref
Cadherins			
e.g., E-cadherin (CD324), N-cadherin (CD325), VE-cadherin (CD144), T-cadherin	Homotypic cell-cell adhesion	Epithelial-mesenchymal transition	[20, 21, 58]
Integrins			
Heterodimers of α- and β-subunits, e.g., α2 (CD49b), α3 (CD49c), α6 (CD49f), αv (CD51), β1 (CD29), β3 (CD51)	Cell-cell and cell-extracellular matrix interactions	Tumorigenesis, invasion, migration, metastases, angiogenesis	[31]
Ig superfamily			
I-CAM (CD54)	Homotypic and heterotypic cell-cell interactions (e.g., interacts with leukocyte function-associated antigen)	Immune suppression, metastasis	[46, 59-63]
M-CAM (MUC18, CD146)	(e.g., interacts with heparan sulfate proteoglycan)	Tumour progression, metastasis	[64-66]
AL-CAM (CD166)	(e.g., interacts with CD6)	Primary tumour growth and invasion. Associated with aggressive phenotype and poor prognosis	[67]
V-CAM-1 (CD106)	(e.g., interacts with integrins)	Transendothelial migration, metastasis	[60, 68]
N-CAM (CD171)	(e.g., interacts with FGFR)	Tumour progression, survival	[69, 70]
Carcinoembryonic antigen (CEA)	(e.g., interacts with CD66a)	Tumorigenesis, invasion, metastasis	[71-73]
Basigin (CD147)	(e.g., interacts with MMPs, N-CAM)	Tumour progression, invasion	[74-80]
OX-2 membrane glycoprotein (CD200)	Immune cell-endothelium interactions	Immune suppression, metastasis	[17-19, 81, 82]
Selectins			
e.g., E-selectin (CD62E), P-selectin (CD62P), L-selectin (CD62L)	Immune cell-endothelium interactions	Organ-specific metastases	[59, 83, 84]
Cell surface proteoglycans			
CD44	Cell-cell and cell-extracellular matrix interactions	Tumour proliferation, migration, invasion, metastasis	[34-39]

Table. 1. (Continued)

Cell surface molecules	Type of interaction	Role in cancer	Ref
Syndecan-1 (CD138)		Migration, invasion, metastasis	[85, 86]
Tyrosine kinase receptor family			
Human epidermal growth factor receptors e.g., HER-2,3,4 (erbB2,3,4)	Heteromolecular interactions, ligand binding activates tyrosine protein kinase signaling activity. Regulates tumour cell interactions with the microenvironment	Tumorigenesis, tumour proliferation, migration, invasion, angiogenesis, organ- specific metastasis	[87]
Eph receptors e.g., EphA2			[88, 89]
C-Met (Hepatocyte growth factor receptors)			[90-92]
Vascular endothelial growth factors receptors e.g., VEGFR1-3			[93-96]
Fibroblast growth factor receptors eg., FGFR1-4 (CD331-4)			[97, 98]

3. Profiling the Plasma Membrane Proteome

A key advantage of proteomics is the ability to identify and quantify proteins on a large scale. The overall goal is to obtain a snapshot of protein expression or abundance levels in a cell, tissue or fluid at a given time. Proteomic analyses of total cell or crude tissue lysates generally identify only a handful of membrane proteins although the plasma membrane proteins are predicted to constitute one third of the human genome [99, 100]. The amphiphilic nature, heterogeneity and relative low abundance of cell surface proteins likely contributes to an under representation of this important sub-proteome in global proteome analyses [101].

There are many different types of proteins associated with the plasma membrane, i.e., integral membrane proteins that transverse the plasma membrane (single-pass type I, type II or multi-pass), proteins attached through a lipid or glycosylphosphatidylinositol (GPI) anchor, and peripheral or membrane-associated proteins [102]. The plasma membrane also contains domains of distinct morphology, composition and function, including lipid rafts, caveolae and tetraspanin-web networks, which may have specialised functions [103]. Moreover, membrane proteins are frequently modified by glycosylation, phosphorylation or sulfonation [102]. It is therefore not surprising that different cell surface protein enrichment methods may yield quite different membrane proteomes, favouring proteins with particular chemistries, interactions or abundance.

The protein composition of the plasma membrane is dynamic, changing with cell differentiation, activation, cell or pathogen contact, microenvironment signalling and during malignant transformation [104]. As encountered in the purification of any organellar structure, absolute purity will never be obtained due to the dynamic nature of cells and their organelles, constantly shuffling proteins between subcellular compartments [105, 106]. Contamination with cytosolic and cytoskeletal proteins or intracellular membranes [103] is common and highly abundant contaminating proteins can obscure the detection of low abundance membrane proteins in the sample [100]. Moreover, plasma membrane proteins are similar to those of other membranous structures of the cell, making their enrichment very difficult [106]. The enrichment of plasma membrane proteins relative to other cellular proteins is important and fractionation techniques must be assessed for the quantity of sample required, reproducibility, protein yield and coverage to ensure a comprehensive analysis.

3.1. Enriching Plasma Membrane Proteins

Membrane enrichment using ultracentrifugation

Combinations of centrifugation protocols are a mainstay of subcellular fractionation. Ultracentrifugation techniques rely on differences in buoyancy density along with size and shape of cellular organelles and can be divided into 2 general categories: differential and density-gradient centrifugation [106-108]. Differential centrifugation primarily separates cellular fractions based on size, whereas density-gradient centrifugation employs a continuous or discontinuous density gradient, isolating different subcellular fractions based on size and density. Plasma membrane proteins are similar to proteins of other membranous structures in the cell, making their enrichment by ultracentrifugation difficult [106]. Due to the low purity of plasma membrane proteins isolated, ultracentrifugation is often only used as an initial step, requiring additional, more selective enrichment methods to increase the purity of isolated membrane proteins.

Chemical proteomics by cell surface labelling

Plasma membrane proteins can be enriched using reactive biotin derivatives that selectively label surface-exposed proteins of intact cells [109]. Biotin is usually coupled to the ε-amino group of lysine using an *N*-hydroxysuccinimide (NHS) ester of a biotin analogue. To selectively label extracellular lysine residues, water-soluble but lipid bilayer impermeable biotin reagents are utilised in a short pulse with quenching to prevent entry of reactive biotin into the cell [5]. Biotinylated proteins can then be extracted from cell lysates using avidin-agarose affinity chromatography. The efficacy of this protocol is dependent on the isolation of intact cells, as disrupted membranes would allow biotin molecules to label intracellular proteins.

Microdomains and microparticle enrichment

The electrostatic attachment of cationic colloidal silica particles to the negatively-charged plasma membrane may yield reasonable cell surface protein enrichments [110]. A subset of plasma membrane proteins may be analyzed by a microparticle protocol that induces the release of membrane fragments from cells [111]. Methods to isolate specific plasma membrane domains such as lipid rafts and caveolae, may exploit the highly insoluble, cholesterol and glycosphingolipid rich features of these regions through differential detergent solubility and sucrose gradient centrifugation [103].

N-linked glycoprotein enrichment using lectin affinity or hydrazide-coupling

Plasma membrane proteins can also be enriched by targeting post-translational modifications that are common to this compartment, such as N-linked glycosylation of the extracellular domains of cell surface proteins [112]. Glycosylation is a post-translational modification involving the addition of polysaccharides to proteins and lipids as an enzyme-directed site-specific process, and is distinct from the non-enzymatic chemical reaction of glycation. Two types of glycosylation exist: *N*-linked glycosylation to the amide nitrogen of asparagine side chains and *O*-linked glycosylation at the hydroxy oxygen of serine and threonine. The majority of plasma membrane proteins are glycosylated, with 80% of the cell surface carbohydrate in *N*-linked glycoprotein form [113]. The surface of human cells and the surrounding extracellular matrix are covered with N-linked polysaccharides, but this modification is less common on intracellular proteins. Protein glycosylation has important roles in growth, migration, differentiation, tumour invasion, host-pathogen interactions, transmembrane signalling, cell-cell interactions and cell communication [114]. Complex carbohydrates at the cell surface appear to serve many functions, such as membrane protein folding and stabilization, antigen determinants (ABO group antigens), cell surface receptor modulation, cell recognition and adhesion [115].

Aberrant glycosylation is a consistent feature of cancer cells, with certain types of glycan structures recognised as well-known markers for neoplastic cell transformation, tumour progression and metastatic potential [116, 117]. The site of glycan attachment to the protein, as well as the type or structure of the polysaccharide, can be modified by malignancy [118]. Therefore, the identification of proteins that are differentially glycosylated between normal and cancerous cells may enable the development of therapeutic antibodies that selectively target transformed cells. Whether such glycosylations result from initial oncogenic transformations or are involved with induction of invasion and metastasis remains controversial [119].

Lectins are proteins that bind non-covalently to certain polysaccharides enabling affinity chromatography of glycoproteins [120]. Because of their unique specificities, lectins have been routinely used as binding tools for a number of applications in glycobiology. On binding to a cell surface, lectins can change the physiology of the plasma membrane causing agglutination and other cell-cell interactions, increase motility, cytoplasmic or nuclear swelling and mitosis [121]. Lectins have also been used for glycoprotein marker detection in numerous cancers [122-126].

Alternatively, glycoproteins can be isolated by oxidizing the polysaccharides to form dialdehydes that are covalently coupled to hydrazide beads for separation via chemical affinity chromatography. PNGase F (peptide-N-glycosidase F) is an enzyme with broad specificity for N-linked sugars, cleaving the link between the polysaccharide and the asparagine residue of the polypeptide, to release peptides for profiling. This method selectively enriches for N-linked glycoproteins or peptides with high yields and minimal contamination [26,27].

3.2. General Proteomics Methodology for Analysing the Plasma Membrane Proteome

2D gel electrophoresis (2DE)

2DE is a high-resolution method for separating proteins according to their isoelectric point (pI) in the first dimension and their molecular weight in the second dimension. Due to their hydrophobicity, 2DE of membrane proteins is challenging. Non-ionic or weak zwitterionic detergents are required for isoelectric focussing (IEF), as opposed to strong ionic detergents that have greater solubilizing power [127]. Hydrophobic proteins don't readily solubilize and may precipitate at their isoelectric point, which corresponds to their lowest point of solubility [128]. Although the development of new zwitterionic detergents (e.g., C7bZO) and improved chaotropic agents (e.g., thiourea) has enhanced gel-based separation of membrane proteins, gel-free protocols generally result in a better representation of the membrane proteome [129].

Mass spectrometry (MS), "slice and dice" and high performance liquid chromatography (HPLC)

MS uses mass analysis for protein characterisation and is the most comprehensive and versatile tool in large-scale proteomics [130]. Tandem mass spectrometry (MS/MS) enables identification of proteins by sequencing their peptides. In the first mass analyser, single peptides are selected and dissociated by collision with an inert gas, such as argon or nitrogen. The resulting fragments are then separated in the second mass analyser, producing the tandem mass or MS/MS spectrum [131]. These fragments are then matched with fragments calculated from peptides in protein databases, enabling protein identification.

The use of sodium dodecyl sulfate polyacrylamide gel electrophoresis (SDS-PAGE) in conjunction with liquid chromatography and MS (LC-MS) has resulted in successful identification of hydrophobic membrane proteins [132]. Transmembrane proteins are efficiently solubilised in an SDS-containing buffer and separated by 1D electrophoresis. The resulting gel is "sliced" into protein bands and each band is "diced" into smaller pieces and the proteins are extracted and digested into peptides. The peptide mixture is then separated based on hydrophobicity by reverse phase (RP) chromatography and injected into the MS for analysis. Using the "slice and dice" method, i.e., SDS-PAGE fractionation coupled with LC-MS/MS analysis, several cell surface proteins associated with tumour metastasis in colorectal cancer have been isolated [133]. Although this approach has improved membrane proteome analyses and is applicable to many of the plasma membrane protein enrichment protocols, normalisation between samples for quantitative studies is problematic.

More commonly, HPLC is utilised for fractionating or simplifying membrane proteomes. LC-MS/MS analysis is faster and less labour intensive than 2DE. HPLC based separations using micro and nanobore columns are scalable to analyse large numbers of biological samples [134] and provides an immediate and comprehensive view of the sample's protein composition. Moreover, for reasons outlined above, LC-MS/MS provides more efficient recovery of membrane proteins in comparison to 2DE. A compelling example of this was the detection of neuroblast differentiation-associated protein AHNAK by LC-MS/MS analysis; a 312 kDa membrane protein highly abundant in melanoma cells that was undetected on 2D gels [134]. Although simple protein mixtures can be analysed by combining LC and MS, complex mixtures overwhelm the resolution and capacity of any 1D chromatography system [135]. 2D–LC separation methods use the independent physical properties of charge and hydrophobicity to resolve complex peptide mixtures before MS [136]. In a typical 2D LC-MS/MS experiment, the acidified complex peptide mixture is applied to a strong cation exchange (SCX) column, and a discrete fraction of the peptides are displaced onto a reverse-phase (RP) chromatography column using a salt step gradient. Peptides are retained on the RP column, but contaminating salts and buffers are removed. The peptides are then eluted from the RP column into the MS using a gradient of increasing acetonitrile. This 2D LC-MS/MS approach has significantly improved the resolution of complex peptide mixtures enabling the identification of cell surface proteins [137].

Comparative proteomics

Quantitative measurements of protein abundance represent one of the key components toward reconstructing a functional cellular network. The introduction of stable isotope labelling by amino acids in cell culture (SILAC), isotope-coded affinity tags (ICAT) [138] and isobaric tags for relative and absolute quantification (iTRAQ™) [139] have greatly facilitated the relative quantification of proteins across multiple samples. In this approach, a denatured, reduced and alkylated protein mixture from each biological sample is digested to generate a mixture of peptides that are then labelled with different tags. The resulting labelled peptide samples are combined and the peptides are separated and analysed using LC-MS/MS. The ratios between the different labels may then be quantified to determine the relative abundance of the labelled peptide between different biological samples. Combining iTRAQ-based labelling and 2D LC-MS/MS allowed the identification of 1460 proteins in membrane-enriched fractions of malignant gliomas [140].

3.3. Cell Surface Proteomics in Leukaemia Research

The characterization of the surface proteome of leukemia cells may contribute to an improved understanding of leukemogenesis, cell signaling, survival mechanisms and therapeutic targets. Recent publications reporting investigations of the plasma membrane proteome of leukemia cells have used several techniques for the recovery and analysis of membrane proteins. Although these studies have led to the identification of several putative biomarkers and drug targets, membrane proteome analysis of leukaemia and lymphoma has not achieved its full potential, and application of these techniques may advance leukemia research.

Boyd *et al* used sucrose-gradient centrifugation to isolate plasma membrane fractions containing membrane proteins and lipid raft micro-domains from mantle cell lymphoma (MCL), identifying several transmembrane, membrane-associated and lipid raft proteins by LC-MS/MS [141]. However, only 25% of the proteins identified had transmembrane regions, and many membrane-associated and contaminating cytosolic proteins were present in the sucrose fractions. Nevertheless, several potential MCL biomarkers from the plasma membrane, lipid rafts and membrane-associated proteins were identified as potential targets for novel therapeutics [141]. Membrane microparticles generated from samples obtained from patients with chronic lymphocytic leukaemia (CLL), small lymphocytic lymphoma (SLL) or MCL were analysed using a "slice and dice" method. Although cytosolic protein contamination was present, several membrane proteins were identified, including many CD antigens. One of these, CD148 was validated as a potential biomarker to distinguish MCL from other mature B-cell neoplasms by flow cytometry [142]. Surfaces of human B- and T-cell lines were analyzed following N-linked glycoprotein enrichment by hydrazide-coupling, identifying many plasma membrane proteins with minimal cytosolic contamination [143]. This method can also be used to establish sites of glycosylation, which is of particular interest in cancer research because glycosylation can be altered by malignancy [118].

3.4. Cell Surface Proteomics in Melanoma Research

Cell surface proteome analyses of primary and metastatic melanoma have great potential to identify markers of tumor progression, metastasis, prognostic outcome as well as new therapeutic targets. Yet, no analyses of clinical melanoma specimens are described in the literature at the time of this publication, despite the promise of this approach in cell line models. Following cell surface biotinylation and affinity purification, Qiu *et al.,* identified more than 100 membrane and membrane-associated proteins from SILAC-labeled primary and metastatic melanoma cells by LC-MS/MS, including a number of surface markers previously associated with tumour progression and metastasis, e.g., adhesion molecules Mel-CAM (CD146) and AL-CAM (CD166), several integrins including $\alpha v\beta 3$, and TKR, HER-2 [144]. Baruthio *et al.,* performed LC-MS/MS analyses of isolated detergent-resistant lipid rafts by sucrose gradient centrifugation from melanoma cell lines derived from tumours at different stages of progression. The enriched membrane sub-proteomes of aggressive, invasive and metastatic cells compared with less malignant, primary melanoma tumour cells showed groups of proteins preferentially associated with the degree of malignancy and possibly implicated in melanoma progression. For instance, decreased expression of the adhesion molecule T-cadherin, previously linked with increased cell proliferation in several cancers, was detected only in the least malignant melanoma cell line [134]. The application of cell surface enrichment techniques described in this chapter, coupled with the powerful separating power of 2D LC-MS/MS has great promise to advance current understanding of melanoma pathophysiology as well as identify new markers and therapeutics.

4. Immunophenotyping Cancers Using an Antibody Microarray

The immunophenotype of the cell is the repertoire of molecules found on the outer surface of the plasma membrane. These molecules have a variety of functions including receptors, adhesion, signalling and transport of molecules in and out of the cell. The combination of cell surface molecules differs according to the function of the cell and it is through these molecules that cells interact with each other. Early research into cell surface molecules produced large numbers of antibodies recognising various epitopes of the same surface molecules. International workshops on Human Leukocyte Differentiation Antigens (HLDA; now known as Human Cell Differentiation Molecules – HCDM) were convened to collect and characterise antibodies against leukocyte and epithelial antigens and then 'cluster' those recognising the same antigens. Those antigens for which at least 2 antibodies have been developed, and that satisfy other criteria, are designated a Cluster of Differentiation (CD) number. Different antibodies against the same CD molecules may recognise different epitopes on the molecule, thus CD molecules have subgroups such as CD1a-e and CD66a-d to denote different isoforms of the same protein. There have been 9 such workshops to date and more than 350 CD antigens defined.

The aim of antibody microarray-based oncoproteomics is to identify novel protein signatures that discriminate between different malignancies. Another potentially important application of surface profiling is associating certain protein signatures with disease progression and survival [3]. In contrast to most antibody microarrays that detect soluble proteins, the DotScan™ antibody microarray (Medsaic Pty. Ltd., Australia) enables the screening of live cell suspensions for the expression of large numbers of cell surface CD antigens in a single assay. Data obtained correlate well with antigen levels determined by flow cytometry [145-147]. DotScan™ has been applied to a variety of clinical samples, including leukaemia, lymphoma, peripheral blood leukocytes from HIV and heart transplant patients, colorectal cancer specimens and metastatic melanoma [147-153]. The use of an extensive immunophenotype (partial membrane proteome) from an antibody microarray may facilitate disease sub-classification and predict the likely clinical behavior of the cancer (e.g., likely sites of metastatic spread), treatment response and patient outcome. Surface antigen expression patterns should also correlate with the mutated genetic program of malignant cells [154].

4.1. Surface Profiling Leukaemias Using an Antibody Microarray

Leukaemias are currently classified using multiple criteria including morphology, karyotype, cytochemistry and a limited immunophenotype of 10-15 CD antigens [155]. Cell surface profiling of CD antigens using the DotScan™ antibody microarray, for example, diagnose acute lymphoblastic leukaemia (ALL) with a certainty of ~95% [146]. Antibodies for lineage-specific antigens capture live cells providing surface profiles that distinguish ALL from normal mononuclear leukocytes and acute myeloid leukaemia (AML) [155]. AML shows stronger expression of the myeloid antigens CD13, CD15 and CD33, the stem cell markers CD34 and CD117 and a range of other antigens [156]. The expression profile obtained using DotScan™ antibody microarray for AML subtype M3 corresponds well with published immunophenotypes obtained by flow cytometry, showing little or no expression of CD11b, CD11c, CD15, CD34, CD45RO, CD86 and HLA-DR, and over-expression of CD9 [146]. Examples of dot binding patterns or immunophenotypes of different leukaemias are shown in Figure 1.

In addition, the DotScan™ CD antibody microarray has been used to profile leukaemias during differentiation and cytotoxic therapy. Treatment of the human myeloid leukaemia cell lines HL60 and NB4 with the differentiation agents all-trans retinoic acid (ATRA), 1α,25-dihydroxyvitamin D_3 and 12-*O*-tetradecanoyl phorbol-13-acetate (TPA), induced some CD antigens, while others were repressed [148, 152]. The anticancer drugs, cladribine and fludarabine, changed the levels of CD antigens on B-lymphoproliferative disorders. For human Raji lymphoma cells, cladribine up-regulated CD10, CD54, CD80 and CD86, with repression of CD22, while fludarabine up-regulated CD20, CD54, CD80, CD86 and CD95. For MEC2 derived from CLL cells, cladribine up-regulated CD11a, CD20, CD43, CD45, CD52, CD54, CD62L CD80, CD86 and CD95, but fludarabine had no effect [157]. Up-regulation of particular CD antigens induced on a leukemia or lymphoma by a chemotherapeutic agent could provide targets for combined treatment with a therapeutic antibody with more selective toxicity.

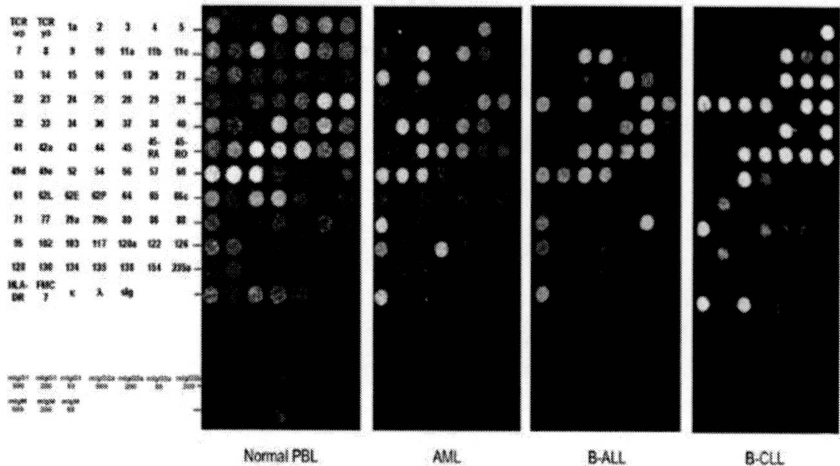

Figure 1. Dot patterns obtained from mononuclear leukocytes captured on DotScan™ microarrays for normal peripheral blood leukocytes (PBL) and samples from patients with acute myeloid leukaemia (AML), acute lymphoblastic leukaemia (ALL) and chronic lymphocytic leukaemia (CLL). The numbers in the key refer to antibodies against the corresponding CD antigens; mIgG1, mIgG2a, mIgG2b and mIgM are murine isotype control antibodies tested at the concentrations shown in μg/mL; TCR α/β, TCR γ/δ, HLA-DR, FMC7, k and λ are antibodies against T cell receptors α/β and γ/δ, HLA-DR, FMC-7, kappa and lambda immunoglobulin light chains, respectively.

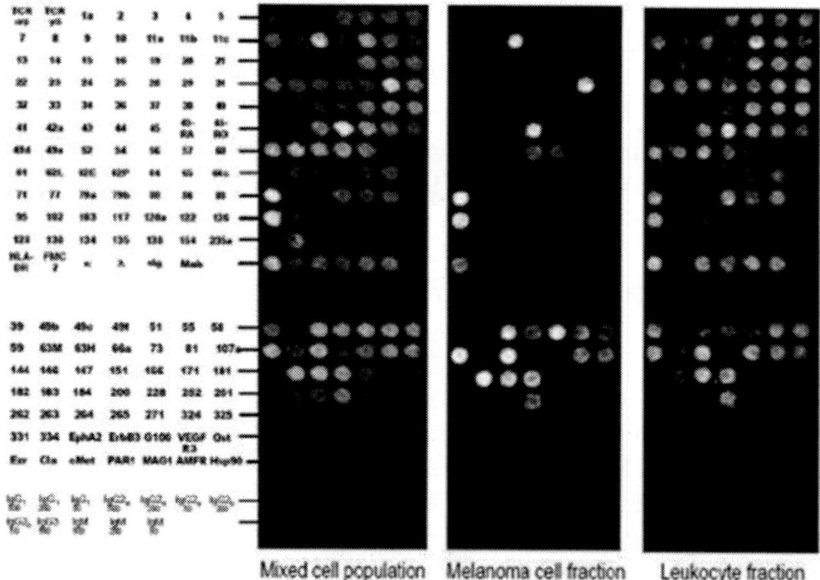

Figure 2. Immunophenotypes of a total metastatic melanoma lymph node sample and cell subsets. From left to right: antibody key and binding patterns of a total cell population, CD45 immuno-depleted melanoma cell fraction and CD45 positive leukocyte fraction. The numbers in the antibody key refer to antibodies against corresponding CD antigens; IgG1, IgG2a, IgG2b, IgG3 and IgM are murine isotype control antibodies tested at the concentrations shown in μg/mL; TCR α/β, TCR γ/δ, HLA-DR, FMC7, k and λ are antibodies against T cell receptors α/β and γ/δ, HLA-DR, FMC-7, kappa and lambda immunoglobulin light chains, respectively.

In a clinical analysis involving 733 patients with a variety of leukaemias and lymphomas and 63 healthy control subjects, Belov *et al.,* demonstrated that surface profiles of CD antigens obtained from a DotScan™ microarray containing 82 monoclonal antibodies against a range of different CD antigens, are sufficient to classify leukaemias from samples of peripheral blood or bone marrow aspirates. The classification of peripheral blood samples using DotScan™ showed a 93.9% correlation with established criteria, and a 96.7% correlation for bone marrow aspirates. These results show that an extensive immunophenotype alone was sufficient to classify the disease when the leukaemic clone dominates the cell population [146).

4.2. Surface Profiling of Metastatic Melanoma Using an Antibody Microarray

Although originally designed for use with leukocytes, the DotScan™ microarray has recently been applied to solid tumour specimens [153, 154, 158]. Capture of live suspended cells from fresh isolates of surgically excised human melanoma on DotScan™ microarrays may provide surface antigen signatures that correlate with disease subtypes, prognosis and drug susceptibility. Cell suspensions made from resected melanoma lymph node metastases contain a variety of cells, including cancer cells, leukocytes (T- and B-lymphocytes, natural killer cells and monocytes), and to a lesser extent, lymph node stromal cells and endothelial cells. Using CD45 antibody-coated magnetic beads, melanoma cells and leukocytes can be enriched from suspensions and profiled separately (see Figure 2). In a pilot study, Kaufman *et al.,* described optimal tissue disaggregation and cell binding conditions to obtain immunophenotypes of enriched metastatic melanoma cell and leukocyte fractions from surgically excised human melanoma metastases from lymph nodes. The differential expression of CD antigens observed for different samples of melanoma cells correlated well with those previously linked to melanoma progression and metastasis, i.e., CD9 [159, 160], CD10 [161, 162], CD13 [163], CD36 [164], CD49d [165], CD54 [166], CD61 [166, 167], CD95 [168] and to poor prognosis, i.e., CD40 [169], CD44 [44, 45] and CD57 [170, 171]. With the procedure established, immunophenotypes of enriched fractions of melanoma cell and leukocytes from lymph node metastases will be determined for a large cohort of Stage III melanoma patients [153].

5. The Therapeutic Potential of Cell Surface Oncoproteomics

Proteomic technologies are being developed to detect cancer earlier, to discover the next generation of targets and biomarkers, and design rational therapeutics according to the molecular profile of the cancer cell, thus tailoring therapies to the patient. The early detection of cancer dramatically reduces mortality. An effective, clinically useful biomarker should be measurable in a readily accessible body fluid, e.g., blood, urine or saliva [172]. Several sub-cellular proteomes, including the ‘secretome’, can be derived from the plasma membrane proteome, as surface proteins can be shed or secreted from the cell as a consequence of oncogenic processes. Proteins released from the cell membrane of malignant cells have been observed in the plasma of patients and the secretome of cancer cells has been a useful source of biomarkers [173, 174]. Therefore, cell surface oncoproteomic analyses may inform plasma-profiling studies, the levels of these proteins in blood could enable diagnosis of disease and severity. For instance, measurement of the cell surface proteins CEA and HER-2 in nipple aspirates has been approved for breast cancer diagnosis in Japan [175]. However, protein patterns or signatures provide greater certainty for diagnosis than a single biomarker [176]. Combining newly discovered biomarkers for ovarian cancer progression with cancer antigen 125 (CA125), has resulted in a more sensitive multi-marker assay that could be useful for screening and suitable for disease monitoring during and after therapy [177]. The DotScan™ microarray with 82 CD antibodies could provide early diagnosis and stratification of haematological malignancies using samples from peripheral blood or bone marrow aspirates [146].

Cytotoxic drugs have been a mainstay of cancer treatment for more than 50 years; however, they show little discrimination between tumour and host cells and may have severe side effects that limit treatment duration, dose and efficacy. [96]. Surface proteins that are expressed selectively on a subset of cancer cells are potential targets for therapeutic agents. For example, monoclonal antibodies (MAbs) directed against growth factor receptors (e.g., anti-HER2, tratuzumab and anti-EGFR, cetuximab) and small molecule inhibitors of tyrosine kinases associated with a particular growth factor receptor (e.g., the c-kit kinase inhibitor, imantinib, and the EGFR kinase inhibitor, NSC56452) [96, 178]. Receptor tyrosine kinases in neoplastic cells may be inhibited by MAbs or kinase inhibitors inducing arrest of cell growth and apoptosis [179, 180]. Plasma membrane biomarkers identified by proteome analysis, such as proteins that are specific to a cancer cell subtype or differentially modified protein isoforms that are not expressed by normal cells, may provide new targets for drug therapy [181].

The use of single drugs to treat cancer often results in resistance and relapse. For example, MAbs generally bind to one surface protein; cancers can acquire mutations that provide resistance, such as reduced expression of the target antigen [182]. Combination therapy reduces the probability of drug resistance and can combat the multi-faceted nature of cancer signalling and survival. Moreover, the use of synergistic therapeutics can result in improved efficacy, eliminating cancer cells before mutations occur, or controlling pro-survival mechanisms that are aberrantly used by malignant cells to evade death. The detection of minimal residual disease (MRD) after therapy is important to avoid relapse or development of resistance, as well as to establish further treatment protocols. Membrane oncoproteomics may provide markers to detect MRD, enabling monitoring of remission status and detection of relapse [183]. The identification of cell surface proteins that predict drug sensitivity will enable tailoring of anti-cancer treatment to the patient with improved clinical outcomes.

Conclusions

Cell surface profiling is of special interest in oncoproteomics as it may provide more accurate diagnosis and prognosis, and identify new therapeutic targets. However, plasma membrane proteomics is challenging, with no ‘gold standard’ technique providing a comprehensive analysis. Discovery proteomics projects, for example, glycoprotein-targeted LC-MS/MS approaches, may provide potential biomarkers that can be screened on large clinical cohorts to determine protein signatures indicative of disease state, stage and outcome. Oncoproteomics will not be limited to the identification of biomarkers for diagnosis and targeted therapy; treatment strategies will be rationally designed to the molecular profile of the cancer, leading to personalised medicine.

References

[1] Ponta, H; Sleeman, J; Herrlich, P. Tumor metastasis formation: cell-surface proteins confer metastasis-promoting or -suppressing properties. *Biochim Biophys Acta*, 1994, May 27, 1198(1), 1-10.

[2] Roesli, C; Borgia, B; Schliemann, C; Gunthert, M; Wunderli-Allenspach, H; Giavazzi, R; et al. Comparative analysis of the membrane proteome of closely related metastatic and nonmetastatic tumor cells. *Cancer Res.*, 2009, Jul 1, 69(13), 5406-14.

[3] Borrebaeck, CA. Antibody microarray-based oncoproteomics. *Expert Opin Biol Ther.*, 2006, Aug, 6(8), 833-8.

[4] Hanahan, D; Weinberg, RA. The hallmarks of cancer. *Cell*, 2000, Jan 7, 100(1), 57-70.

[5] Jang, JH; Hanash, S. Profiling of the cell surface proteome. *Proteomics*, 2003, Oct, 3(10), 1947-54.

[6] Lund, R; Leth-Larsen, R; Jensen, ON; Ditzel, HJ. Efficient isolation and quantitative proteomic analysis of cancer cell plasma membrane proteins for identification of metastasis-associated cell surface markers. *J Proteome Res.*, 2009, Jun, 8(6), 3078-90.

[7] Zbytek, B; Carlson, JA; Granese, J; Ross, J; Mihm, MC; Slominski, A. Current concepts of metastasis in melanoma. *Expert Rev Dermatol*, 2008, Oct, 3(5), 569-85.

[8] Groh, V; Wu, J; Yee, C; Spies, T. Tumour-derived soluble MIC ligands impair expression of NKG2D and T-cell activation. *Nature*, 2002, Oct 17, 419(6908), 734-8.

[9] Caligiuri, MA. Immune surveillance against common cancers: the great escape. *Blood*, 2005, 106(3), 773-4.

[10] Bauer, S; Groh, V; Wu, J; Steinle, A; Phillips, JH; Lanier, LL; et al. Activation of NK cells and T cells by NKG2D, a receptor for stress-inducible MICA. *Science*, 1999, Jul 30, 285(5428), 727-9.

[11] Wright, GJ; Cherwinski, H; Foster-Cuevas, M; Brooke, G; Puklavec, MJ; Bigler, M; et al. Characterization of the CD200 receptor family in mice and humans and their interactions with CD200. *J Immunol*, 2003, Sep 15, 171(6), 3034-46.

[12] Hoek, RM; Ruuls, SR; Murphy, CA; Wright, GJ; Goddard, R; Zurawski, SM; et al. Down-regulation of the macrophage lineage through interaction with OX2 (CD200). *Science*, 2000, Dec 1, 290(5497), 1768-71.

[13] Jenmalm, MC; Cherwinski, H; Bowman, EP; Phillips, JH; Sedgwick, JD. Regulation of myeloid cell function through the CD200 receptor. *J Immunol*, 2006, Jan 1, 176(1), 191-9.

[14] Gorczynski, L; Chen, Z; Hu, J; Kai, Y; Lei, J; Ramakrishna, V; et al. Evidence that an OX-2-positive cell can inhibit the stimulation of type 1 cytokine production by bone

marrow-derived B7-1 (and B7-2)-positive dendritic cells. *J Immunol*, 1999, Jan 15, 162(2), 774-81.

[15] Gorczynski, RM; Lee, L; Boudakov, I. Augmented Induction of CD4+CD25+ Treg using monoclonal antibodies to CD200R. *Transplantation*, 2005, May 15, 79(9), 1180-3.

[16] Curiel, TJ; Coukos, G; Zou, L; Alvarez, X; Cheng, P; Mottram, P; et al. Specific recruitment of regulatory T cells in ovarian carcinoma fosters immune privilege and predicts reduced survival. *Nat Med*, 2004, Sep, 10(9), 942-9.

[17] Petermann, KB; Rozenberg, GI; Zedek, D; Groben, P; McKinnon, K; Buehler, C; et al. CD200 is induced by ERK and is a potential therapeutic target in melanoma. *J Clin Invest*, 2007, Dec, 117(12), 3922-9.

[18] McWhirter, JR; Kretz-Rommel, A; Saven, A; Maruyama, T; Potter, KN; Mockridge, CI; et al. Antibodies selected from combinatorial libraries block a tumor antigen that plays a key role in immunomodulation. *Proc Natl Acad Sci U S A*, 2006, Jan 24, 103(4), 1041-6.

[19] Kretz-Rommel, A; Qin, F; Dakappagari, N; Ravey, EP; McWhirter, J; Oltean, D; et al. CD200 expression on tumor cells suppresses antitumor immunity: new approaches to cancer immunotherapy. *J Immunol*, 2007, May 1, 178(9), 5595-605.

[20] Perez-Moreno, M; Jamora, C; Fuchs, E. Sticky business: orchestrating cellular signals at adherens junctions. *Cell*, 2003, Feb 21, 112(4), 535-48.

[21] Conacci-Sorrell, M; Zhurinsky, J; Ben-Ze'ev, A. The cadherin-catenin adhesion system in signaling and cancer. *J Clin Invest*, 2002, Apr, 109(8), 987-91.

[22] Wheelock, MJ; Johnson, KR. Cadherins as modulators of cellular phenotype. *Annu Rev Cell Dev Biol.*, 2003, 19, 207-35.

[23] Bolos, V; Peinado, H; Perez-Moreno, MA; Fraga, MF; Esteller, M; Cano, A. The transcription factor Slug represses E-cadherin expression and induces epithelial to mesenchymal transitions: a comparison with Snail and E47 repressors. *J Cell Sci.*, 2003, Feb 1, 116(Pt 3), 499-511.

[24] Cano, A; Perez-Moreno, MA; Rodrigo, I; Locascio, A; Blanco, MJ; del Barrio, MG; et al. The transcription factor snail controls epithelial-mesenchymal transitions by repressing E-cadherin expression. *Nat Cell Biol.*, 2000, Feb, 2(2), 76-83.

[25] Peinado, H; Ballestar, E; Esteller, M; Cano, A. Snail mediates E-cadherin repression by the recruitment of the Sin3A/histone deacetylase 1 (HDAC1)/HDAC2 complex. *Mol Cell Biol.*, 2004, Jan, 24(1), 306-19.

[26] Perez-Moreno, MA; Locascio, A; Rodrigo, I; Dhondt, G; Portillo, F; Nieto, MA; et al. A new role for E12/E47 in the repression of E-cadherin expression and epithelial-mesenchymal transitions. *J Biol Chem.*, 2001, Jul 20, 276(29), 27424-31.

[27] Yang, J; Mani, SA; Donaher, JL; Ramaswamy, S; Itzykson, RA; Come, C; et al. Twist, a master regulator of morphogenesis, plays an essential role in tumor metastasis. *Cell*, 2004, Jun 25, 117(7), 927-39.

[28] Graff, JR; Gabrielson, E; Fujii, H; Baylin, SB; Herman, JG. Methylation patterns of the E-cadherin 5' CpG island are unstable and reflect the dynamic, heterogeneous loss of E-cadherin expression during metastatic progression. *J Biol Chem.*, 2000, Jan 28, 275(4), 2727-32.

[29] Nass, SJ; Herman, JG; Gabrielson, E; Iversen, PW; Parl, FF; Davidson, NE; et al. Aberrant methylation of the estrogen receptor and E-cadherin 5' CpG islands increases

with malignant progression in human breast cancer. *Cancer Res.*, 2000, Aug 15, 60(16), 4346-8.

[30] Bacac, M; Stamenkovic, I. Metastatic cancer cell. *Annu Rev Pathol*, 2008, 3, 221-47.

[31] Hood, JD; Cheresh, DA. Role of integrins in cell invasion and migration. *Nat Rev Cancer*, 2002, Feb, 2(2), 91-100.

[32] Mizejewski, GJ. Role of integrins in cancer: survey of expression patterns. *Proc Soc Exp Biol Med*, 1999, Nov, 222(2), 124-38.

[33] Brooks, PC; Clark, RA; Cheresh, DA. Requirement of vascular integrin alpha v beta 3 for angiogenesis. *Science*, 1994, Apr 22, 264(5158), 569-71.

[34] Kosaki, R; Watanabe, K; Yamaguchi, Y. Overproduction of hyaluronan by expression of the hyaluronan synthase Has2 enhances anchorage-independent growth and tumorigenicity. *Cancer Res.*, 1999, Mar 1, 59(5), 1141-5.

[35] Naor, D; Sionov, RV; Ish-Shalom, D. CD44: structure, function, and association with the malignant process. *Adv Cancer Res.*, 1997, 71, 241-319.

[36] Pauli, BU; Knudson, W. Tumor invasion: a consequence of destructive and compositional matrix alterations. *Hum Pathol*, 1988, Jun, 19(6), 628-39.

[37] Thomas, L; Etoh, T; Stamenkovic, I; Mihm, MC; Jr., Byers, HR. Migration of human melanoma cells on hyaluronate is related to CD44 expression. *J Invest Dermatol*, 1993, Feb, 100(2), 115-20.

[38] Knudson, CB; Knudson, W. Hyaluronan-binding proteins in development, tissue homeostasis, and disease. *FASEB J*, 1993, Oct, 7(13), 1233-41.

[39] Zhang, L; Underhill, CB; Chen L. Hyaluronan on the surface of tumor cells is correlated with metastatic behavior. *Cancer Res.*, 1995, Jan 15, 55(2), 428-33.

[40] van Muijen, GN; Danen, EH; Veerkamp, JH; Ruiter, DJ; Lesley, J; van den Heuvel, LP. Glycoconjugate profile and CD44 expression in human melanoma cell lines with different metastatic capacity. *Int J Cancer*, 1995, Apr 10, 61(2), 241-8.

[41] Gardner, MJ; Catterall, JB; Jones, LM; Turner, GA. Human ovarian tumour cells can bind hyaluronic acid via membrane CD44: a possible step in peritoneal metastasis. *Clin Exp Metastasis*, 1996, Sep, 14(4), 325-34.

[42] Birch, M; Mitchell, S; Hart, IR. Isolation and characterization of human melanoma cell variants expressing high and low levels of CD44. *Cancer Res.*, 1991, Dec 15, 51(24), 6660-7.

[43] Cannistra, SA; Kansas, GS; Niloff, J; DeFranzo, B; Kim, Y; Ottensmeier, C. Binding of ovarian cancer cells to peritoneal mesothelium in vitro is partly mediated by CD44H. *Cancer Res.*, 1993, Aug 15, 53(16), 3830-8.

[44] Karjalainen, JM; Tammi, RH; Tammi, MI; Eskelinen, MJ; Agren, UM; Parkkinen, JJ; et al. Reduced level of CD44 and hyaluronan associated with unfavorable prognosis in clinical stage I cutaneous melanoma. *Am J Pathol*, 2000, Sep, 157(3), 957-65.

[45] Dietrich, A; Tanczos, E; Vanscheidt, W; Schopf, E; Simon, JC. High CD44 surface expression on primary tumours of malignant melanoma correlates with increased metastatic risk and reduced survival. *Eur J Cancer*, 1997 May, 33(6), 926-30.

[46] Horst, E; Meijer, CJ; Radaszkiewicz, T; Ossekoppele, GJ; Van Krieken, JH; Pals, ST. Adhesion molecules in the prognosis of diffuse large-cell lymphoma: expression of a lymphocyte homing receptor (CD44), LFA-1 (CD11a/18), and ICAM-1 (CD54). *Leukemia*, 1990, Aug, 4(8), 595-9.

[47] Horst, E; Meijer, CJ; Radaskiewicz, T; van Dongen, JJ; Pieters, R; Figdor, CG; et al. Expression of a human homing receptor (CD44) in lymphoid malignancies and related stages of lymphoid development. *Leukemia*, 1990, May, 4(5), 383-9.

[48] Pals, ST; Horst, E; Ossekoppele, GJ; Figdor, CG; Scheper, RJ; Meijer, CJ. Expression of lymphocyte homing receptor as a mechanism of dissemination in non-Hodgkin's lymphoma. *Blood*, 1989, Mar, 73(4), 885-8.

[49] Sviatoha, V; Rundgren, A; Tani, E; Hansson, J; Kleina, R; Skoog, L. Expression of CD40, CD44, bcl-2 antigens and rate of cell proliferation on fine needle aspirates from metastatic melanoma. *Cytopathology*, 2002, Feb, 13(1), 11-21.

[50] Hsieh, HF; Yu, JC; Ho, LI; Chiu, SC; Harn, HJ. Molecular studies into the role of CD44 variants in metastasis in gastric cancer. *Mol Pathol*, 1999, Feb, 52(1), 25-8.

[51] Edlund, M; Sung, SY; Chung, LW. Modulation of prostate cancer growth in bone microenvironments. *J Cell Biochem*, 2004, Mar 1, 91(4), 686-705.

[52] Triozzi, PL; Eng, C; Singh, AD. Targeted therapy for uveal melanoma. *Cancer Treat Rev.*, 2008, May, 34(3), 247-58.

[53] Nguyen, DX; Bos, PD; Massague, J. Metastasis: from dissemination to organ-specific colonization. *Nat Rev Cancer*, 2009 Apr, 9(4), 274-84.

[54] Burstein, HJ; Lieberman, G; Slamon, DJ; Winer, EP; Klein, P. Isolated central nervous system metastases in patients with HER2-overexpressing advanced breast cancer treated with first-line trastuzumab-based therapy. *Ann Oncol*, 2005, Nov, 16(11), 1772-7.

[55] Abdel-Ghany, M; Cheng, HC; Elble, RC; Pauli, BU. The breast cancer beta 4 integrin and endothelial human CLCA2 mediate lung metastasis. *J Biol Chem.*, 2001, Jul 6, 276(27), 25438-46.

[56] Brown, DM; Ruoslahti, E. Metadherin, a cell surface protein in breast tumors that mediates lung metastasis. *Cancer Cell*, 2004, Apr, 5(4), 365-74.

[57] Kang, Y; Siegel, PM; Shu, W; Drobnjak, M; Kakonen, SM; Cordon-Cardo, C; et al. A multigenic program mediating breast cancer metastasis to bone. *Cancer Cell*, 2003, Jun, 3(6), 537-49.

[58] Bussemakers, MJ; Schalken, JA. The role of cell adhesion molecules and proteases in tumor invasion and metastasis. *World J Urol*, 1996, 14(3), 151-6.

[59] Fox, SB; Turner, GD; Gatter, KC; Harris, AL. The increased expression of adhesion molecules ICAM-3, E- and P-selectins on breast cancer endothelium. *J Pathol*, 1995, Dec, 177(4), 369-76.

[60] Natali, P; Nicotra, MR; Cavaliere, R; Bigotti, A; Romano, G; Temponi, M; et al. Differential expression of intercellular adhesion molecule 1 in primary and metastatic melanoma lesions. *Cancer Res.*, 1990, Feb 15, 50(4), 1271-8.

[61] Roland, CL; Dineen, SP; Toombs, JE; Carbon, JG; Smith, CW; Brekken, RA; et al. Tumor-derived intercellular adhesion molecule-1 mediates tumor-associated leukocyte infiltration in orthotopic pancreatic xenografts. *Exp Biol Med (Maywood)*. 2010, Feb, 235(2), 263-70.

[62] Fujihara, T; Yashiro, M; Inoue, T; Sawada, T; Kato, Y; Ohira, M; et al. Decrease in ICAM-1 expression on gastric cancer cells is correlated with lymph node metastasis. *Gastric Cancer*, 1999, Dec, 2(4), 221-5.

[63] Ren, G; Zhao, X; Zhang, L; Zhang, J; L'Huillier, A; Ling, W; et al. Inflammatory cytokine-induced intercellular adhesion molecule-1 and vascular cell adhesion

molecule-1 in mesenchymal stem cells are critical for immunosuppression. *J Immunol*, 2010, Mar 1, 184(5), 2321-8.

[64] Lehmann, JM; Riethmuller, G; Johnson, JP. MUC18, a marker of tumor progression in human melanoma, shows sequence similarity to the neural cell adhesion molecules of the immunoglobulin superfamily. *Proc Natl Acad Sci U S A*, 1989, Dec, 86(24), 9891-5.

[65] Staquicini, FI; Tandle, A; Libutti, SK; Sun, J; Zigler, M; Bar-Eli, M; et al. A subset of host B lymphocytes controls melanoma metastasis through a melanoma cell adhesion molecule/MUC18-dependent interaction, evidence from mice and humans. *Cancer Res.*, 2008, Oct 15, 68(20), 8419-28.

[66] Shih, IM; Speicher, D; Hsu, MY; Levine, E; Herlyn, M. Melanoma cell-cell interactions are mediated through heterophilic Mel-CAM/ligand adhesion. *Cancer Res.*, 1997, Sep 1, 57(17), 3835-40.

[67] Swart, GW; Lunter, PC; Kilsdonk, JW; Kempen, LC. Activated leukocyte cell adhesion molecule (ALCAM/CD166), signaling at the divide of melanoma cell clustering and cell migration? *Cancer Metastasis Rev.*, 2005, Jun, 24(2), 223-36.

[68] Klemke, M; Weschenfelder, T; Konstandin, MH; Samstag, Y. High affinity interaction of integrin alpha4beta1 (VLA-4) and vascular cell adhesion molecule 1 (VCAM-1) enhances migration of human melanoma cells across activated endothelial cell layers. *J Cell Physiol*, 2007, Aug, 212(2), 368-74.

[69] Roesler, J; Srivatsan, E; Moatamed, F; Peters, J; Livingston, EH. Tumor suppressor activity of neural cell adhesion molecule in colon carcinoma. *Am J Surg.*, 1997, Sep, 174(3), 251-7.

[70] Tezel, E; Kawase, Y; Takeda, S; Oshima, K; Nakao, A. Expression of neural cell adhesion molecule in pancreatic cancer. *Pancreas*, 2001, Mar, 22(2), 122-5.

[71] Obrink, B. CEA adhesion molecules: multifunctional proteins with signal-regulatory properties. *Curr Opin Cell Biol.*, 1997, Oct, 9(5), 616-26.

[72] Sienel, W; Dango, S; Woelfle, U; Morresi-Hauf, A; Wagener, C; Brummer, J; et al. Elevated expression of carcinoembryonic antigen-related cell adhesion molecule 1 promotes progression of non-small cell lung cancer. *Clin Cancer Res.*, 2003, Jun, 9(6), 2260-6.

[73] Singer, BB; Scheffrahn, I; Kammerer, R; Suttorp, N; Ergun, S; Slevogt, H. Deregulation of the CEACAM expression pattern causes undifferentiated cell growth in human lung adenocarcinoma cells. *PLoS One*, 2010, 5(1), e8747.

[74] Kataoka, H; DeCastro, R; Zucker, S; Biswas, C. Tumor cell-derived collagenase-stimulatory factor increases expression of interstitial collagenase, stromelysin, and 72-kDa gelatinase. *Cancer Res.*, 1993, Jul 1, 53(13), 3154-8.

[75] Bordador, LC; Li, X; Toole, B; Chen, B; Regezi, J; Zardi, L; et al. Expression of emmprin by oral squamous cell carcinoma. *Int J Cancer*, 2000, Feb 1, 85(3), 347-52.

[76] Lim, M; Martinez, T; Jablons, D; Cameron, R; Guo, H; Toole, B; et al. Tumor-derived EMMPRIN (extracellular matrix metalloproteinase inducer) stimulates collagenase transcription through MAPK p38. *FEBS Lett*, 1998, Dec 11, 441(1), 88-92.

[77] Muraoka, K; Nabeshima, K; Murayama, T; Biswas, C; Koono, M. Enhanced expression of a tumor-cell-derived collagenase-stimulatory factor in urothelial carcinoma, its usefulness as a tumor marker for bladder cancers. *Int J Cancer*, 1993, Aug 19, 55(1), 19-26.

[78] Polette, M; Gilles, C; Marchand, V; Lorenzato, M; Toole, B; Tournier, JM; et al. Tumor collagenase stimulatory factor (TCSF) expression and localization in human lung and breast cancers. *J Histochem Cytochem*, 1997, May, 45(5), 703-9.

[79] Sameshima, T; Nabeshima, K; Toole, BP; Yokogami, K; Okada, Y; Goya, T; et al. Glioma cell extracellular matrix metalloproteinase inducer (EMMPRIN) (CD147) stimulates production of membrane-type matrix metalloproteinases and activated gelatinase A in co-cultures with brain-derived fibroblasts. *Cancer Lett*, 2000, Sep 1, 157(2), 177-84.

[80] Kanekura, T; Chen, X; Kanzaki, T. Basigin (CD147) is expressed on melanoma cells and induces tumor cell invasion by stimulating production of matrix metalloproteinases by fibroblasts. *Int J Cancer*, 2002, Jun 1, 99(4), 520-8.

[81] Ko, YC; Chien, HF; Jiang-Shieh, YF; Chang, CY; Pai, MH; Huang, JP; et al. Endothelial CD200 is heterogeneously distributed, regulated and involved in immune cell-endothelium interactions. *J Anat*, 2009 Jan, 214(1), 183-95.

[82] Stumpfova, M; Ratner, D; Desciak, EB; Eliezri, YD; Owens, DM. The immunosuppressive surface ligand CD200 augments the metastatic capacity of squamous cell carcinoma. *Cancer Res.*, 2010, Apr 1, 70(7), 2962-72.

[83] Ye, C; Kiriyama, K; Mistuoka, C; Kannagi, R; Ito, K; Watanabe, T; et al. Expression of E-selectin on endothelial cells of small veins in human colorectal cancer. *Int J Cancer*, 1995, May 16, 61(4), 455-60.

[84] Makrilia, N; Kollias, A; Manolopoulos, L; Syrigos, K. Cell adhesion molecules, role and clinical significance in cancer. *Cancer Invest*, 2009, Dec, 27(10), 1023-37.

[85] Kusumoto, T; Kodama, J; Seki, N; Nakamura, K; Hongo, A; Hiramatsu, Y. Clinical significance of syndecan-1 and versican expression in human epithelial ovarian cancer. *Oncol Rep.*, 2010, Apr, 23(4), 917-25.

[86] Ishikawa, T; Kramer, RH. Sdc1 negatively modulates carcinoma cell motility and invasion. *Exp Cell Res.*, 2010, Apr 1, 316(6), 951-65.

[87] Holbro, T; Hynes, NE. ErbB receptors: directing key signaling networks throughout life. *Annu Rev Pharmacol Toxicol*, 2004, 44, 195-217.

[88] Zelinski, DP; Zantek, ND; Stewart, JC; Irizarry, AR; Kinch, MS. EphA2 overexpression causes tumorigenesis of mammary epithelial cells. *Cancer Res.*, 2001, Mar 1, 61(5), 2301-6.

[89] Walker-Daniels, J; Hess, AR; Hendrix, MJ; Kinch, MS. Differential regulation of EphA2 in normal and malignant cells. *Am J Pathol*, 2003, Apr, 162(4), 1037-42.

[90] Cruz, J; Reis-Filho, JS; Silva, P; Lopes, JM. Expression of c-met tyrosine kinase receptor is biologically and prognostically relevant for primary cutaneous malignant melanomas. *Oncology*, 2003, 65(1), 72-82.

[91] Birchmeier, C; Birchmeier, W; Gherardi, E; Vande Woude, GF. Met, metastasis, motility and more. *Nat Rev Mol Cell Biol.*, 2003, Dec, 4(12), 915-25.

[92] Christensen, JG; Burrows, J; Salgia, R. c-Met as a target for human cancer and characterization of inhibitors for therapeutic intervention. *Cancer Lett*, 2005, Jul 8, 225(1), 1-26.

[93] Liu, B; Ma, J; Wang, X; Su, F; Li, X; Yang, S; et al. Lymphangiogenesis and Its Relationship With Lymphatic Metastasis and Prognosis in Malignant Melanoma. *Anat Rec (Hoboken)*. 2008, Jun 16.

[94] Ferrara, N. Vascular endothelial growth factor as a target for anticancer therapy. *Oncologist*, 2004, 9 Suppl 1, 2-10.

[95] Cao, Y. Antiangiogenic cancer therapy. *Semin Cancer Biol.*, 2004, Apr, 14(2), 139-45.

[96] Press, MF; Lenz, HJ. EGFR, HER2 and VEGF pathways: validated targets for cancer treatment. *Drugs.*, 2007, 67(14), 2045-75.

[97] Knights, V; Cook, SJ. De-regulated FGF receptors as therapeutic targets in cancer. *Pharmacol Ther.*, 2010, Jan, 125(1), 105-17.

[98] Jeffers, M; LaRochelle, WJ; Lichenstein, HS. Fibroblast growth factors in cancer: therapeutic possibilities. *Expert Opin Ther Targets.*, 2002, Aug, 6(4), 469-82.

[99] Wu, CC; Yates, JR; 3rd. The application of mass spectrometry to membrane proteomics. *Nat Biotechnol*, 2003, Mar, 21(3), 262-7.

[100] Macher, BA; Yen, TY. Proteins at membrane surfaces-a review of approaches. *Mol Biosyst*, 2007, Oct, 3(10), 705-13.

[101] Helbig, AO; Heck, AJ; Slijper, M. Exploring the membrane proteome--challenges and analytical strategies. *J Proteomics*, 2010, Mar 10, 73(5), 868-78.

[102] Tan, S; Tan, HT; Chung MC. Membrane proteins and membrane proteomics. *Proteomics*, 2008, Oct, 8(19), 3924-32.

[103] Josic, D; Clifton, JG. Mammalian plasma membrane proteomics. *Proteomics*, 2007, Aug, 7(16), 3010-29.

[104] Cordwell, SJ; Thingholm, TE. Technologies for plasma membrane proteomics. *Proteomics*, 2009, Oct 15.

[105] Yates, JR; 3rd, Gilchrist, A; Howell, KE; Bergeron, JJ. Proteomics of organelles and large cellular structures. *Nat Rev Mol Cell Biol.*, 2005, Sep, 6(9), 702-14.

[106] Elschenbroich, S; Kim, Y; Medin, JA; Kislinger, T. Isolation of cell surface proteins for mass spectrometry-based proteomics. *Expert Rev Proteomics*, 2010, Feb, 7(1), 141-54.

[107] Castle, JD. Purification of organelles from mammalian cells. *Curr Protoc Immunol*, 2003, Nov, Chapter 8, Unit 8 1B.

[108] Huber, LA; Pfaller, K; Vietor, I. Organelle proteomics: implications for subcellular fractionation in proteomics. *Circ Res.*, 2003, May 16, 92(9), 962-8.

[109] Shin, BK; Wang, H; Yim, AM; Le Naour, F; Brichory, F; Jang, JH; et al. Global profiling of the cell surface proteome of cancer cells uncovers an abundance of proteins with chaperone function. *J Biol Chem.*, 2003, Feb 28, 278(9), 7607-16.

[110] Rahbar, AM; Fenselau, C. Integration of Jacobson's pellicle method into proteomic strategies for plasma membrane proteins. *J Proteome Res.*, 2004, Nov-Dec, 3(6), 1267-77.

[111] Miguet, L; Bechade, G; Fornecker, L; Zink, E; Felden, C; Gervais, C; et al. Proteomic Analysis of Malignant B-Cell Derived Microparticles Reveals CD148 as a Potentially Useful Antigenic Biomarker for Mantle Cell Lymphoma Diagnosis. *J Proteome Res.*, 2009, Jul 6, 8(7), 3346-54.

[112] Gahmberg, CG; Tolvanen, M. Why mammalian cell surface proteins are glycoproteins. *Trends Biochem Sci.*, 1996, Aug, 21(8), 308-11.

[113] Aarnoudse, CA; Garcia Vallejo JJ; Saeland E; van Kooyk Y. Recognition of tumor glycans by antigen-presenting cells. *Curr Opin Immunol*, 2006, Feb, 18(1), 105-11.

[114] Zhao, YY; Takahashi, M; Gu, JG; Miyoshi, E; Matsumoto, A; Kitazume, S; et al. Functional roles of N-glycans in cell signaling and cell adhesion in cancer. *Cancer Sci.*, 2008, Jul, 99(7), 1304-10.

[115] Brandley, BK; Schnaar, RL. Cell-surface carbohydrates in cell recognition and response. *J Leukoc Biol.*, 1986, Jul, 40(1), 97-111.

[116] Fuster, MM; Esko, JD. The sweet and sour of cancer: glycans as novel therapeutic targets. *Nat Rev Cancer*, 2005, Jul, 5(7), 526-42.

[117] An, HJ; Kronewitter, SR; de Leoz, ML; Lebrilla, CB. Glycomics and disease markers. *Curr Opin Chem Biol.*, 2009 Dec, 13(5-6), 601-7.

[118] Peracaula, R; Barrabes, S; Sarrats, A; Rudd, PM; de Llorens, R. Altered glycosylation in tumours focused to cancer diagnosis. *Dis Markers*, 2008, 25(4-5), 207-18.

[119] Hakomori, S. Glycosylation defining cancer malignancy: new wine in an old bottle. *Proc Natl Acad Sci U S A*, 2002, Aug 6, 99(16), 10231-3.

[120] Wei, X; Li, L. Comparative glycoproteomics: approaches and applications. *Brief Funct Genomic Proteomic*, 2009 Mar, 8(2), 104-13.

[121] Sharon, N; Lis, H. History of lectins: from hemagglutinins to biological recognition molecules. *Glycobiology*, 2004, Nov, 14(11), 53R-62R.

[122] Nakamura, N; Ota, H; Katsuyama, T; Akamatsu, T; Ishihara, K; Kurihara, M; et al. Histochemical reactivity of normal, metaplastic, and neoplastic tissues to alpha-linked N-acetylglucosamine residue-specific monoclonal antibody HIK1083. *J Histochem Cytochem*, 1998, Jul, 46(7), 793-801.

[123] Cochran, AJ; Wen, DR; Berthier-Vergnes, O; Bailly, C; Dore, JF; Berard, F; et al. Cytoplasmic accumulation of peanut agglutinin-binding glycoconjugates in the cells of primary melanoma correlates with clinical outcome. *Hum Pathol*, 1999 May, 30(5), 556-61.

[124] Drake, RR; Schwegler, EE; Malik, G; Diaz, J; Block, T; Mehta, A; et al. Lectin capture strategies combined with mass spectrometry for the discovery of serum glycoprotein biomarkers. *Mol Cell Proteomics*, 2006, Oct, 5(10), 1957-67.

[125] Schwegler, E; Malik, G; Ward, M; Mitchell, S; Diaz, J; Davis, J; et al. Differential lectin capture profiling and identification of serum glycoproteins to detect and monitor prostate cancer. *AACR Meeting Abstracts*, 2006, 675-6.

[126] Zhao, J; Patwa, TH; Qiu, W; Shedden, K; Hinderer, R; Misek, DE; et al. Glycoprotein microarrays with multi-lectin detection: unique lectin binding patterns as a tool for classifying normal, chronic pancreatitis and pancreatic cancer sera. *J Proteome Res.*, 2007, May, 6(5), 1864-74.

[127] Molloy, MP. Two-dimensional electrophoresis of membrane proteins using immobilized pH gradients. *Anal Biochem.*, 2000, Apr 10, 280(1), 1-10.

[128] Adessi, C; Miege, C; Albrieux, C; Rabilloud, T. Two-dimensional electrophoresis of membrane proteins: a current challenge for immobilized pH gradients. *Electrophoresis*, 1997, Jan, 18(1), 127-35.

[129] Washburn, MP; Wolters, D; Yates, JR; 3rd. Large-scale analysis of the yeast proteome by multidimensional protein identification technology. *Nat Biotechnol*, 2001, Mar, 19(3), 242-7.

[130] Yates, JR; Ruse, CI; Nakorchevsky, A. Proteomics by mass spectrometry: approaches, advances, and applications. *Annu Rev Biomed Eng.*, 2009, 11, 49-79.

[131] Mann, M; Hendrickson, RC; Pandey, A. Analysis of proteins and proteomes by mass spectrometry. *Annu Rev Biochem*, 2001, 70, 437-73.

[132] Jungblut, PR; Schiele, F; Zimny-Arndt, U; Ackermann, R; Schmid, M; Lange, S; et al. Helicobacter pylori proteomics by 2-DE/MS, 1-DE-LC/MS and functional data mining. *Proteomics*, Jan, 10(2), 182-93.

[133] Luque-Garcia, JL; Martinez-Torrecuadrada, JL; Epifano, C; Canamero, M; Babel, I; Casal, JI. Differential protein expression on the cell surface of colorectal cancer cells associated to tumor metastasis. *Proteomics*, 2010, Mar, 10(5), 940-52.

[134] Baruthio, F; Quadroni, M; Ruegg, C; Mariotti, A. Proteomic analysis of membrane rafts of melanoma cells identifies protein patterns characteristic of the tumor progression stage. *Proteomics*, 2008, Nov, 8(22), 4733-47.

[135] McCormack, AL; Schieltz, DM; Goode, B; Yang, S; Barnes, G; Drubin, D; et al. Direct analysis and identification of proteins in mixtures by LC/MS/MS and database searching at the low-femtomole level. *Anal Chem.*, 1997, Feb 15, 69(4), 767-76.

[136] Link, AJ; Eng, J; Schieltz, DM; Carmack, E; Mize, GJ; Morris, DR; et al. Direct analysis of protein complexes using mass spectrometry. *Nat Biotechnol*, 1999, Jul, 17(7), 676-82.

[137] Ruth, MC; Old, WM; Emrick, MA; Meyer-Arendt, K; Aveline-Wolf, LD; Pierce, KG; et al. Analysis of membrane proteins from human chronic myelogenous leukemia cells: comparison of extraction methods for multidimensional LC-MS/MS. *J Proteome Res.*, 2006, Mar, 5(3), 709-19.

[138] Gygi, SP; Rist, B; Gerber, SA; Turecek, F; Gelb, MH; Aebersold, R. Quantitative analysis of complex protein mixtures using isotope-coded affinity tags. *Nat Biotechnol*, 1999, Oct, 17(10), 994-9.

[139] Ross, PL; Huang, YN; Marchese, JN; Williamson, B; Parker, K; Hattan, S; et al. Multiplexed protein quantitation in Saccharomyces cerevisiae using amine-reactive isobaric tagging reagents. *Mol Cell Proteomics*, 2004, Dec, 3(12), 1154-69.

[140] Rajcevic, U; Petersen, K; Knol, JC; Loos, M; Bougnaud, S; Klychnikov, O; et al. iTRAQ-based proteomics profiling reveals increased metabolic activity and cellular cross-talk in angiogenic compared with invasive glioblastoma phenotype. *Mol Cell Proteomics*, 2009, Nov, 8(11), 2595-612.

[141] Boyd, RS; Jukes-Jones, R; Walewska, R; Brown, D; Dyer, MJ; Cain, K. Protein profiling of plasma membranes defines aberrant signaling pathways in mantle cell lymphoma. *Mol Cell Proteomics*, 2009, Jul, 8(7), 1501-15.

[142] Miguet, L; Bechade, G; Fornecker, L; Zink, E; Felden, C; Gervais, C; et al. Proteomic analysis of malignant B-cell derived microparticles reveals CD148 as a potentially useful antigenic biomarker for mantle cell lymphoma diagnosis. *J Proteome Res.*, 2009, Jul, 8(7), 3346-54.

[143] Wollscheid, B; Bausch-Fluck, D; Henderson, C; O'Brien, R; Bibel, M; Schiess, R; et al. Mass-spectrometric identification and relative quantification of N-linked cell surface glycoproteins. *Nat Biotechnol*, 2009, Apr, 27(4), 378-86.

[144] Qiu, H; Wang, Y. Quantitative analysis of surface plasma membrane proteins of primary and metastatic melanoma cells. *J Proteome Res.*, 2008, May, 7(5), 1904-15.

[145] Belov, L; de la Vega, O; dos Remedios, CG; Mulligan, SP; Christopherson, RI. Immunophenotyping of leukemias using a cluster of differentiation antibody microarray. *Cancer Res.*, 2001, Jun 1, 61(11), 4483-9.

[146] Belov, L; Mulligan, SP; Barber, N; Woolfson, A; Scott M; Stoner, K; et al. Analysis of human leukaemias and lymphomas using extensive immunophenotypes from an antibody microarray. *Br J Haematol*, 2006, Oct, 135(2), 184-97.

[147] Belov, L; Huang, P; Barber, N; Mulligan, SP; Christopherson, RI. Identification of repertoires of surface antigens on leukemias using an antibody microarray. *Proteomics*, 2003, Nov, 3(11), 2147-54.

[148] White, SL; Belov, L; Barber, N; Hodgkin, PD; Christopherson, RI. Immunophenotypic changes induced on human HL60 leukaemia cells by 1alpha,25-dihydroxyvitamin D3 and 12-O-tetradecanoyl phorbol-13-acetate. *Leuk Res.*, 2005, Oct, 29(10), 1141-51.

[149] Wu, JQ; Wang, B; Belov, L; Chrisp, J; Learmont, J; Dyer, WB; et al. Antibody microarray analysis of cell surface antigens on CD4+ and CD8+ T cells from HIV+ individuals correlates with disease stages. *Retrovirology*, 2007, 4, 83.

[150] Woolfson, A; Stebbing, J; Tom, BD; Stoner, KJ; Gilks, WR; Kreil, DP; et al. Conservation of unique cell-surface CD antigen mosaics in HIV-1-infected individuals. *Blood*, 2005, Aug 1, 106(3), 1003-7.

[151] Lal, S; Lui, R; Nguyen, L; Macdonald, P; Denyer, G; dos Remedios, C. Increases in leukocyte cluster of differentiation antigen expression during cardiopulmonary bypass in patients undergoing heart transplantation. *Proteomics*, 2004, Jul, 4(7), 1918-26.

[152] Barber, N; Belov, L; Christopherson, RI. All-trans retinoic acid induces different immunophenotypic changes on human HL60 and NB4 myeloid leukaemias. *Leuk Res.*, 2008, Feb, 32(2), 315-22.

[153] Kaufman, KL; Belov, L; Huang, P; Mactier, S; Scolyer, RA; Mann, GJ; et al. An extended antibody microarray for surface profiling metastatic melanoma. *J Immunol Methods*, 2010, Apr 2.

[154] Ellmark, P; Belov, L; Huang, P; Lee, CS; Solomon, MJ; Morgan, DK; et al. Multiplex detection of surface molecules on colorectal cancers. *Proteomics*, 2006, Mar, 6(6), 1791-802.

[155] ES. Jaffe, NLH; H. Stei,n, Vardiman, JW. World Health Organization Classification of Tumours: *Pathology and Genetics of Tumours of Haematopoietic and Lymphoid Tissues Lyon*, France: IARC Press, 2001.

[156] Barber, N; Gez, S; Belov, L; Mulligan, SP; Woolfson, A; Christopherson, RI. Profiling CD antigens on leukaemias with an antibody microarray. *FEBS Lett*, 2009, Mar 17.

[157] Cassano, C; Mulligan, SM; S. Belov, L; Huang, P; Christopherson, R. Cladribine and fludarabine nucleoside change the levels of CD antigens on B-lymphoproliferative disorders. *International Journal of Proteomics*, 2010.

[158] Zhou, J; Belov, L; Huang, PY; Shin, JS; Solomon, MJ; Chapuis, PH; et al. Surface antigen profiling of colorectal cancer using antibody microarrays with fluorescence multiplexing. *J Immunol Methods*, 2010, Apr 15, 355(1-2), 40-51.

[159] Si, Z; Hersey, P. Expression of the neuroglandular antigen and analogues in melanoma. CD9 expression appears inversely related to metastatic potential of melanoma. *Int J Cancer*, 1993, Apr 22, 54(1), 37-43.

[160] Hong, IK; Kim, YM; Jeoung, DI; Kim, KC; Lee, H. Tetraspanin CD9 induces MMP-2 expression by activating p38 MAPK, JNK and c-Jun pathways in human melanoma cells. *Exp Mol Med*, 2005, Jun 30, 37(3), 230-9.

[161] Bilalovic, N; Sandstad, B; Golouh, R; Nesland, JM; Selak, I; Torlakovic, EE. CD10 protein expression in tumor and stromal cells of malignant melanoma is associated with tumor progression. *Mod Pathol*, 2004, Oct, 17(10), 1251-8.

[162] Kanitakis, J; Narvaez, D; Claudy, A. Differential expression of the CD10 antigen (neutral endopeptidase) in primary versus metastatic malignant melanomas of the skin. *Melanoma Res.*, 2002, Jun, 12(3), 241-4.

[163] Fujii, H; Nakajima, M; Saiki, I; Yoneda, J; Azuma, I; Tsuruo, T. Human melanoma invasion and metastasis enhancement by high expression of aminopeptidase N/CD13. *Clin Exp Metastasis*, 1995, Sep, 13(5), 337-44.

[164] Thorne, RF; Marshall, JF; Shafren, DR; Gibson, PG; Hart, IR; Burns, GF. The integrins alpha3beta1 and alpha6beta1 physically and functionally associate with CD36 in human melanoma cells. Requirement for the extracellular domain OF CD36. *J Biol Chem.*, 2000, Nov 10, 275(45), 35264-75.

[165] Nikkola, J; Vihinen, P; Vlaykova, T; Hahka-Kemppinen, M; Heino, J; Pyrhonen, S. Integrin chains beta1 and alphav as prognostic factors in human metastatic melanoma. *Melanoma Res.*, 2004, Feb, 14(1), 29-37.

[166] Johnson, JP. Cell adhesion molecules in the development and progression of malignant melanoma. *Cancer Metastasis Rev.*, 1999, 18(3), 345-57.

[167] McGary, EC; Lev, DC; Bar-Eli, M. Cellular adhesion pathways and metastatic potential of human melanoma. *Cancer Biol Ther.*, 2002, Sep-Oct, 1(5), 459-65.

[168] Owen-Schaub, LB; van Golen, KL; Hill, LL; Price, JE. Fas and Fas ligand interactions suppress melanoma lung metastasis. *J Exp Med*, 1998, Nov 2, 188(9), 1717-23.

[169] van den Oord, JJ; Maes, A; Stas, M; Nuyts, J; Battocchio, S; Kasran, A; et al. CD40 is a prognostic marker in primary cutaneous malignant melanoma. *Am J Pathol*, 1996, Dec, 149(6), 1953-61.

[170] Casado, JG; Delgado, E; Patsavoudi, E; Duran, E; Sanchez-Correa, B; Morgado, S; et al. Functional implications of HNK-1 expression on invasive behaviour of melanoma cells. *Tumour Biol.*, 2008, 29(5), 304-10.

[171] Thies, A; Schachner, M; Berger, J; Moll, I; Schulze, HJ; Brunner, G; et al. The developmentally regulated neural crest-associated glycotope HNK-1 predicts metastasis in cutaneous malignant melanoma. *J Pathol*, 2004, Aug, 203(4), 933-9.

[172] Petricoin, EF; Zoon, KC; Kohn, EC; Barrett, JC; Liotta, LA. Clinical proteomics: translating benchside promise into bedside reality. *Nat Rev Drug Discov*, 2002, Sep, 1(9), 683-95.

[173] Schiess, R; Wollscheid, B; Aebersold, R. Targeted proteomic strategy for clinical biomarker discovery. *Mol Oncol*, 2009, Feb, 3(1), 33-44.

[174] Hanash, SM; Pitteri, SJ; Faca, VM. Mining the plasma proteome for cancer biomarkers. *Nature*, 2008, Apr 3, 452(7187), 571-9.

[175] Kurebayashi, J. Biomarkers in breast cancer. *Gan To Kagaku Ryoho*, 2004, Jul, 31(7), 1021-6.

[176] Jain, KK. Innovations, challenges and future prospects of oncoproteomics. *Mol Oncol*, 2008, Aug, 2(2), 153-60.

[177] Helleman, J; van der Vlies, D; Jansen, MP; Luider, TM; van der Burg, ME; Stoter, G; et al. Serum proteomic patterns for ovarian cancer monitoring. *Int J Gynecol Cancer*, 2008, Sep-Oct, 18(5), 985-95.

[178] Yang, RY; Yang, KS; Pike, LJ; Marshall, GR. Targeting the Dimerization of Epidermal Growth Factor Receptors with Small-Molecule Inhibitors. *Chem Biol Drug Des.*, 2010, May 4, Epub.
[179] Sierra, JR; Cepero, V; Giordano, S. Molecular mechanisms of acquired resistance to tyrosine kinase targeted therapy. *Mol Cancer*, 2010, 9, 75.
[180] Cepero, V; Sierra, JR; Giordano, S. Tyrosine Kinases as Molecular Targets to Inhibit Cancer Progression and Metastasis. *Curr Pharm Des.*, 2010, Feb 18, Epub.
[181] Kohnke, PL; Mulligan, SP; Christopherson, RI. Membrane proteomics for leukemia classification and drug target identification. *Curr Opin Mol Ther.*, 2009, Dec, 11(6), 603-10.
[182] Petrelli, A; Valabrega, G. Multitarget drugs: the present and the future of cancer therapy. *Expert Opin Pharmacother*, 2009, Mar, 10(4), 589-600.
[183] Campana, D. Molecular determinants of treatment response in acute lymphoblastic leukemia. *Hematology Am Soc Hematol Educ Program*, 2008, 366-73.

In: Oncoproteins: Types and Detection
Editor: Jeremy R. Davis
ISBN: 978-1-61761-551-1

Chapter 4

Finding for New Site Specific References for Oncoprotein Tumor Markers: An Example from Thai Context

Viroj Wiwanitkit
Wiwanitkit House, Bangkhae, Bangkok Thailand

Abstract

There are several presently used oncoprotein tumor markers. These markers are very useful in oncology and widely used worldwide. An important problem in interpretation of tumor marker result is the reference. The site specific references are needed for better interpretation of tumor maker result. In this specific paper, the author will present the approach on finding for new site specific references for oncoprotein tumor markers. An example from Thai context will be used as a model.

Introduction

There are several presently used oncoprotein tumor markers. These markers are very useful in oncology and widely used worldwide. An important problem in interpretation of tumor marker result is the reference. The site specific references are needed for better interpretation of tumor maker result. To get this specific purpose, there are many steps to be fulfilled. First, the adjustment of the reference value should be performed. This should be firstly in universal or global view. There is no need to perform the adjustment for specific site reference value. Second, the consideration for using of the marker comparing with other alternatives as well as using combining with other alternative should also be done. At this stage, the cost effectiveness analysis might be performed to get the site specific most cost effective alternative node. The final derived alternative can be set as new site specific reference method.

In this specific paper, the author will present the approach on finding for new site specific references for oncoprotein tumor markers. An example from Thai context will be used as a model. The scenario of CA-125, an oncoprotein which is widely used as a tumor marker for ovarian cancer will be hereby demonstrated.

Step 1: Adjustment to Get New Reference Value

Model: Adjusted CA – 125 Cut off Level For Screening For Ovarian Cancer

1. Introduction

CA – 125 is a kind of tumor marker which is useful in early diagnosis of ovarian cancer [1 - 5]. Similar to other tumor markers, CA – 125 is a serum test that is useful in preventive oncology [1 – 5]. Since ovarian cancer is a common cancer in female, screening by serum CA – 125 determination becomes a useful method in cancer prevention. An important problems in using CA – 125 is its diagnostic property. According to the literature, the diagnostic sensitivity for CA - 125 ranges from 70 % in premenopause subjects to 94 % in menopause subjects. However, the difference in cut off level must be appliedfor diagnosis (> 200 U/mL for premenopause subject and > 35 U/mL for menopause subject). To solve the problem of diagnostic property, a new diagnostic tool namely risk of malignancy index (RMI) was recently proposed [6 – 7]. The RMI s the application of ultrasonography (USG) examination to basic CA -125 test. The RMI is the result from three parameters, ultrasound score, menopause stage score and CA-125 level. The cut off level for RMI, not similar to classical CA -125, has a single cut off level at 200 [6 – 7]. This cut off level is applicable for both premenopaue and menopause cases or any age groups. However, a big problem for using RMI in the developing countries can be seen. The problem is mainly due to the availability and affordability of USG examination. Here, the author tries to set up a new adjusted CA -125 cut off level based on RMI to make the CA -125 more useful in screening for ovarian cancer.

2. Materials and Methods

This is a basic study to develop a new mathematical model. RMI was used as a primary model for developing adjusted CA-125 cut off level. Basically, as already noted, RMI [6 – 7] is equal to "USG score x menopausal stage score X CA-125 level". USG score is equal to 1 if there is only one lesion and equal to 3 if there are more than one lesions from USG examinations. Menopausal stage score is equal to 1 for premenopause cases and equal to 3 for menopausal cases. The adjusted CA-125 is directly calculated at the standard cut off level of RMI, 200.

3. Results

The new cut ff level of CA-125 at each classification of subject are calculated and presented in Table 1. The range of new derived CA-125 level from calculation is between

22.2 and 200 U/mL. For getting the best sensitivity, the cut off value at 22.2 is selected to be the best new adjusted CA-125 cut off level.

Table 1. A table of categorized new CA 125 levels (U/mL) in different classification of the patients

Menopausal stage	Number of lesion		
	Know		Not know
	1	> 1	
Premenopause	200	66.7	133.6
Menopause	66.7	22.2	44.5

4. Discussion

CA -125 is a good useful blood test for screening for ovarian cancer. It can be used in primary setting; the blood test can be performed on the referred clotted blood sample at the standard reference laboratory [1 – 5]. This is automatically done by automatic analyzer at present. This is easier than the use of USG machine which requires expert in the field for performing the test and interpret the result. Recently, a new RMI, a combined CA-125 determination and USG test was proposed [6 – 7]. RMI brings better diagnostic sensitivity with one cut off level in screening for ovarian cancer. However, its main obstacle is the limitation in using of USG machine. The author hereby report on the adjusted CA-125 cut off level developed based on the RMI model. The new adjusted CA-125 cut off level is finally derived. In addition, a table of categorized new CA 125 levels in different classification of the patients is also presented. These data is useful in increasing diagnostic property in screening for ovarian cancer by simple CA-125 determination.

Step 2: Consideration on Comparing as well as Combining with Other Alternatives and Further Site Specific Cost Effectiveness Analysis

Model: Combination among Symptom Index, CA125, and HE4 as Alternative Tool for PREDIction of Ovarian Cancer: A Cost Effectiveness Analysis

1. Introduction

Ovarian cancer is a big problem in gynecological oncology. An early detection of this oncological disorder seems to be an effective tool in preventive oncology. The screening for ovarian cancer is widely studied. Among several tools for screening, the three screening alternatives, symptom index (SI) [8 - 9], CA125 [10], and human epididymis protein 4 (HE4) [11] are commonly used for prediction of ovarian cancer. Recently, Anderson et al studied on the diagnostic property of these three tests and proposed that a combination of these tests would add value in diagnosis but did not reported on the cost effectiveness of those tests [8].

Here, the author performed an additional cost effectiveness analysis on the combination among symptom index, CA125, and HE4 as alternative tool for prediction of ovarian cancer.

2. Materials and Methods

This study is a descriptive study based on cost effectiveness analysis. The three tools for prediction of ovarian cancer, SI, CA125 and HE4, were assessed. The cost in this study was referred to the standard cost at the reference laboratory in Bangkok Thailand (Special Laboratory). The effectiveness of each tool and combination among the three tests (SI and CA125, SI and HE4, SI and CA125and HE4) was reviewed. The data from the previous validated studies were used [8 – 12]. The final cost effectiveness was done to find the cost per effectiveness of each alternative. This technique of cost effectiveness study is the standard method that is used in many previous publications in laboratory medicine [13 - 17].

3. Result

The cost of each tool was presented in Table 2. The effectiveness of each tool and combination among the three tests (SI and CA125, SI and HE4, SI and CA125and HE4) was presented in the same table. The cost per effectiveness of SI is lowest and that of SI-CA125-HE4 is the highest. The details of cost effectiveness analysis are also shown in Table 2.

4. Discussion

There are new attempts to develop diagnostic tool to predict ovarian cancer. Classically, the SI is used. It is not surprised that this alternative has the lowest cost per effectiveness but SI poses very poor sensitivity and cannot be accepted for clinical practice. The low sensitivity implies very high false diagnosis [8 – 9]. The advent in biomarkers can help correct the problem of diagnosis of ovarian cancer. Some biomarkers are mentioned for its effectiveness in diagnosis of ovarian cancer. The classical one is CA125 [18]. However, this tumor marker is still not the ideal marker for ovarian cancer [17]. Recently, another new marker, HE4 is proposed but it is still not an ideal one [19]. Li et al proposed that "Preliminary data show that HE4 may have more potential than cancer antigen 125 in discriminating benign from cancerous ovarian masses, and has the strongest correlation with endometrial cancer of all markers tested to date [19]."

Table 2. Cost, effectiveness and cost effectiveness of the studied alternative

Alternative	Cost (US dollar)	Effectiveness (%)[1]	Cost-effectiveness (US dollar)
SI	1.5	64	2.3
CA125	14.3	79.4	18.0
HE4	42.9	80.4	53.4
SI-CA125	15.8	89.3	17.7
SI-HE4	44.4	84	52.9
HE4-CA125	57.2	93.8	61
SI-CA125-SE4	58.7	95	61.8

[1]Effectiveness in this work is set as sensitivity for prediction based on average data from previous validated reports [8 – 12].

The combination of several tools becomes a new concept to increase the diagnostic property but an important concern is the increased cost for screening. Although there are some reports confirming the increased sensitivity [8 – 12] but there is no report on cost effectiveness. Here, the author performs an appraisal on this topic. It is no doubt that any combinations in this study shows increased diagnostic property. However, the increased cost can also be seen. Of several combinations, the combination that poses the lowest cost per effectiveness is SI-CA125. Its cost effectiveness is significant lower than other combinations. Hence, it is hereby suggested that this combination should be selected in clinical practice

References

[1] Bast, RC; Jr; Hunter, V; Knapp, RC. *Pros and cons of gynecologic tumor Cancer.* 1987, Oct 15, 60(8 Suppl), 1984-92.

[2] Jacobs, I; Bast, RC. Jr. The CA 125 tumour-associated antigen *Hum Reprod*, 1989, Jan, 4(1), 1-12.

[3] Pabst, T; Ludwig, C. CA 125--a tumor *Schweiz Med Wochenschr*. 1995 Jun 17, 125(24), 1195-200.

[4] Göcze, P; Vahrson, H. Ovarian carcinoma. *Orv Hetil*, 1993, Apr 25, 134(17), 915-8.

[5] Pennehouat, G; Gugliemina, JN; Naouri, M; Créquat, J; Bouret, JM; Thébault, Y; Madelenat, P. Screening for ovarian cancer. *Contracept Fertil Sex*, 1993, Mar, 21(3), 223-30.

[6] Geomini, P; Kruitwagen, R; Bremer, GL; Cnossen, J; Mol, BW. The accuracy *Obstet Gynecol*, 2009, Feb, 113(2 Pt 1), 384-94.

[7] Bailey, J; Tailor, A; Naik, R; Lopes, A; Godfrey, K; Hatem, HM; Monaghan, J. Risk of malignancy *Int J Gynecol Cancer*, 2006, Jan-Feb, 16 Suppl 1, 30-4.

[8] Andersen, MR; Goff, BA; Lowe, KA; Scholler, N; Bergan, L; Dresher, CW; Paley, P; Urban, N. Combining a symptoms *Cancer*, 2008, Aug 1, 113(3), 484-9.

[9] Andersen, MR; Goff, BA; Lowe, KA; Scholler, N; Bergan, L; Drescher, CW; Paley, P; Urban, N. Use of a Symptom Index, CA125, and HE4 to predict ovarian cancer. *Gynecol Oncol.* 2009, Nov 27. [Epub ahead of print]

[10] Bast, RC; Jr; Badgwell, D; Lu, Z; Marquez, R; Rosen, D; Liu, J; Baggerly, KA; Atkinson, EN; Skates, S; Zhang, Z; Lokshin, A; Menon, U; Jacobs, I; Lu, K. New tumor markers: CA125 and beyond. *Int J Gynecol Cancer.* 2005, Nov-Dec, 15 Suppl, 3, 274-81.

[11] Moore, RG; McMeekin, DS; Brown, AK; DiSilvestro, P; Miller, MC; Allard, WJ; Gajewski, W; Kurman, R; Bast, RC; Jr; Skates, SJ. A novel multiple marker bioassay *Gynecol Oncol*, 2009, Jan, 112(1), 40-6.

[12] Shah, CA; Lowe, KA; Paley, P; Wallace, E; Anderson, GL; McIntosh, MW; Andersen, MR; Scholler, N; Bergan, LA; Thorpe, JD; Urban, N; Drescher, CW. Influence of ovarian cancer risk status on the diagnostic performance of the serum biomarkers mesothelin, HE4, and CA125. *Cancer Epidemiol Biomarkers Prev.* 2009, May, 18(5), 1365-72.

[13] Wiwanitkit, V. Screening for cervical cancer *Asian Pac J Cancer Prev.*, 2009, Jul-Sep, 10(3), 531-2.

[14] Wiwanitkit, V. Screening for syphilis *Arch Gynecol Obstet*, 2007, Dec, 276(6), 629-31.
[15] Wiwanitkit, V. Study of the cost *Trop Doct*, 2005, Jan, 35(1), 23-5.
[16] Asswawitoontip, S; Wiwanitkit, V. Cost-effective study of determination methods for low-density lipoprotein *J Med Assoc Thai.*, 2002, Jun, 85 Suppl 1, S91-6.
[17] Suwansaksri, J; Wiwanitkit, V; Paritpokee, N. Screening for hemoglobin *Arch Gynecol Obstet*, 2004, Dec, 270(4), 211-3.
[18] Bouanène, H; Miled, A. Tumor Marker CA125: biochemical and molecular properties. *Bull Cancer*, 2009, May, 96(5), 597-601.
[19] Li, J; Dowdy, S; Tipton, T; Podratz, K; Lu, WG; Xie, X; Jiang, SW. HE4 as a biomarker for ovarian and endometrial cancer *Expert Rev Mol Diagn*, 2009, Sep, 9(6), 555-66.

In: Oncoproteins: Types and Detection
Editor: Jeremy R. Davis

ISBN: 978-1-61761-551-1

Chapter 5

Oncogenic Activity of Epstein-Barr Virus Encoded Latent Membrane Protein 1 in Undifferentiated Nasopharyngeal Carcinoma

Thian-Sze Wong, Jimmy Yu-Wai Chan, Victor Shing-Howe To, Raymond King-Yin Tsang, Wai-Kuen Ho and William Ignace Wei
Department of Surgery, The University of Hong Kong

Abstract

Epstein-Barr virus (EBV) is a ubiquitous herpesvirus. Epstein-Barr virus (EBV)-encoded latent membrane protein 1 (LMP1) expression is closely associated with the malignant transformation of undifferentiated nasophayngeal carcinoma (NPC). LMP1 plays a strong role in activating the anti-apoptotic response in the host cells. It could also stimulate growth and division of the infected nasopharyngeal epithelial cells. Further, LMP1 expression is associated with the aggressiveness of the disease. Recent studies revealed that LMP1 could alter the host's epigenetic machinery and activate expression of oncogenic microRNAs. Thus, LMP1 is considered as the key mediator in transforming normal nasopharyngeal epithelial cells. In terms of regulation, expression of LMP1 is regulated largely by epigenetic mechanisms such as CpG island methylation. However, the key triggering factors for LMP1 upregulation is not yet found. Recent study indicated that the infected EBV is involved in controlling LMP1 levels. EBV-encoded microRNA could bind to the 3' UTR of the LMP1 transcript and hence hindering the translation of LMP1 protein. Understanding the regulation dynamics of LMP1 is essential in developing novel prevention or treatment regime for the disease.

Introduction

Undifferentiated nasopharyngeal carcinoma (NPC) is a squamous cell carcinoma arising from the nasopharyngeal epithelium. The cancer cells are shaped as large polygonal cells with oval or round nucleus containing scanty chromatin and distinct nucleoli. The tumor is non-lymphomatous but containing prominent lymphocytic infiltration [1]. The disease is first report in 1901 and is considered as a rare malignancy [2]. The global incidence is about 1 per 100,000 individuals [3]. However, the incidence is particularly higher in southern China, northern Africa, and Alaska [1]. Genetic, viral, and environmental factors would all contribute to the aetiology of the disease [4]. In particular, NPC from regions with either high or low incidence rate are infected with Epstein-Barr virus (EBV). EBV is a prototype gamma herpes virus and is the first virus, which shows to have a close association with the development of human cancers [5]. EBV can immortalize B-cells *in vitro* and is used as a mean to establish lymphoblastoid cell lines [6]. In NPC patients, EBV-specific antibody is consistently detected in the serum and is considered as a molecular indicator to monitor the disease [4]. In addition, EBV-encoded DNA, RNA and protein is detected in NPC cells suggesting a close association of EBV-infection and pathogenesis of the disease [4]. According to WHO classification, undifferentiated NPC belongs to type III non-keratinisng carcinoma and is much more radiosensitive than the type I keartinising squamous cell carcinoma.

EBV and LMP1

EBV establishes latent infection in all population and is asymptomatic. Primary infection usually occurs within the first decades of life through contact with infected saliva. The circular double-stranded EBV episome persists in the infected cells and is replicated by DNA polymerase in the nuclei [7]. Bornkamm *et al* classified the life cycle of EBV into 3 phases: (1) maintaining the episomal state; (2) establishment of latency; and (3) reactivation, replication, and synthesis of progeny [8]. In undifferentiated NPC, EBV is in type II latency and is characterized by expression of particular set of viral genes such as BamHI A RNA transcripts (BART), BARF, EBER, EBNA, LMP1 and LMP2 [9, 10].

EBV encoded several transforming proteins and LMP is one of the major types. The EBV genome encoded 3 LMP (LMP1, 2A, and 2B), of which, LMP1 is most potent in transforming cancer cells. The oncogenic function of LMP1 is first demonstrated by its ability to transform rodent-fibroblast *in vitro* [11]. LMP1 gene encoded an integral membrane protein (386 amino acid) and is transcribed leftwards in EBV genome. Structurally, LMP1 protein consists of an internal hydrophilic N-terminus, hydrophobic transmembrane domains and internal cytoplasmic domain [8]. Functionally, it is a constitutively active tumor necrosis factor receptor [12]. Further, it could upregulate anti-apoptotic proteins in the host cells [13]. The precise mechanisms leading to LMP1 activation is not clear. It is now know that LMP1 is regulated by epigenetic mechanisms and the promoter region of LMP1 gene is regulated by CpG methylation [14]. The following discussion will try to link up the current knowledge of LMP1 and the epigenetic mechanisms of NPC.

Epigenetics of Host Genome

Epigenetic alterations refer to alterations in gene expression that are not caused by changes in DNA sequences [15]. The alterations may or may not heritable, albeit it will ultimately change the transcriptional potential of the cells [16). In general, epigenetic regulations of gene expression in eukaryotic cells are carried out by (a) cytosine methylation; (b) histone modification and chromatin remodeling; and (c) microRNA expression/ suppression. In cancer cells, dysregulation of the epigenetic mechanism is a common event. In virus associated cancer such as NPC, increasing evidence suggesting that the oncoproteins encoded by the viral genome cold play a part in altering the normal epigenetic mechanisms leading to the aberrant gene expression in cancer cells.

DNA Methylaton and LMP1

Aberrant DNA methylation is a common epigenetic alteration of undifferentiated nasopharyngeal carcinoma [17, 18]. Methylation of the CpG islands located within the gene regulatory regions would hinder gene expression through transcription interference. Recently, it was demonstrated that EBV in the infected NPC has the ability to silence tumor suppressor genes in the host genome and is partly done by LMP1 upregulation. In NPC cell line, LMP1 can activates DNA methyltransferase 1, 3a, and 3b leading to hypermethylation of key tumor suppressor genes in undifferentiated NPC including E-cadherin and retinoic receptor-beta 2 [19 - 21].

Microrna of the Host Cell and LMP1

It has long been know that LMP1 encoded by the EBV has a regulatory role to the expression of host's genes. However, the precise interactions between LMP1 protein and transcription machinery of the host are still missing. Recent findings suggested that LMP1 didn't suppress or interfere with the tumor suppressor gene transcripts or protein directly. It did so by triggering particular microRNAs which would subsequently degrade the transcripts and/or hinder the protein translation process. For instance, LMP1 could induce the host cell to express cellular microRNA miR-155, an oncogenic microRNA identified in nasopharyngeal carcinoma and several human malignancies [22]. Although the precise mechanism is still unclear, increasing data suggested that this triggering effect is achieved by activating nuclear factor-kappaB (NF-kappaB) in the host cells [23].

Viral Microrna and LMP1

Apart from altering the host's microRNA expression patterns, the EBV itself could encode viral microRNA. EBV is the first human virus identified to have the ability to express viral microRNA [9]. EBV-encoded microRNA could target cellular genes directly. For example, EBV microRNA BHRF1-3 could target he IFN-inducible T-cell attracting chemokine CXCL-11/I-TAC and hence suppress its expression [24]. Further, expression of viral microRNAs is employed by the latently infected EBV to regulate the expression of LMP1. BamHI-A rightward transcripts (BARTs) encoded microRNA, for example, is shown to target the 3′ UTR of the LMP1 gene and thereby hindering its expression [25]. Expression of EBV-encoded BART microRNA is now recognized as a post-transcriptional regulation mechanism in controlling LMP1 levels in the host cells. As LMP1 could control the expression of key cancer-related genes in the host cells, EBV could control the host's gens exprssion patterns indirectly with EBV-encoded microRNA.

Conclusion

Although much of the works has been done on EBV LMP1, how LMP1 could tightly regulate NPC pathogenesis in every stages of NPC is not well understood. As mentioned above, the role of LMP1 is multifaceted. Interplays between EBV-encoded microRNA, cellular expressed microRNA and LMP1 would have a determining role on the fate of the infected cells. Further studies are warranted to display the whole picture of LMP1 in the pathogenesis of undifferentiated NPC.

References

[1] Wei, WI; Sham, JS. Nasopharyngeal carcinoma. *Lancet*, 2005, Jun 11-17, 365(9476), 2041-54.

[2] Jackson, C. Primary carcinoma of the nasopharynx: a table of cases. *J Am Med Assoc*, 1901, Aug 10, 37(6), 371-377.

[3] Yu, MC; Yuan, JM. Epidemiology of nasopharyngeal carcinoma. *Semin Cancer Biol.*, 2002, Dec, 12(6), 421-9.

[4] Razak, AR; Siu, LL; Liu, FF; Ito, E; O'Sullivan, B; Chan, K. Nasopharyngeal carcinoma: The next challenges. *Eur J Cancer*, 2010, May 5.

[5] Epstein, MA; Achong, BG; Barr, YM; Zajac, B; Henle, G; Henle, W. Morphological and virological investigations on cultured Burkitt tumor lymphoblasts (strain Raji). *J Natl Cancer Inst.*, 1966, Oct, 37(4), 547-59.

[6] Young, LS; Rickinson, AB. Epstein-Barr virus: 40 years on. *Nat Rev Cancer*, 2004, Oct, 4(10), 757-68.

[7] Adams, A. Replication of latent Epstein-Barr virus genomes in Raji cells. *J Virol*, 1987, May, 61(5), 1743-6.

[8] Bornkamm, GW; Hammerschmidt, W. Molecular virology of Epstein-Barr virus. *Philos Trans R Soc Lond B Biol Sci.*, 2001, Apr 29, 356(1408), 437-59.

[9] Lo, AK; Lo, KW; Tsao, SW; Wong, HL; Hui, JW; To, KF; Hayward, DS; Chui, YL; Lau, YL; Takada, K; Huang, DP. Epstein-Barr virus infection alters cellular signal cascades in human nasopharyngeal epithelial cells. *Neoplasia*, 2006, Mar, 8(3), 173-80.

[10] Tsao, SW; Tramoutanis, G; Dawson, CW; Lo, AK; Huang, DP. The significance of LMP1 expression in nasopharyngeal carcinoma. *Semin Cancer Biol.*, 2002, Dec, 12(6), 473-87.

[11] Wang, D; Liebowitz, D; Kieff, E. An EBV membrane protein expressed in immortalized lymphocytes transforms established rodent cells. *Cell*, 1985, Dec, 43(3 Pt 2), 831-40.

[12] Uchida, J; Yasui, T; Takaoka-Shichijo, Y; Muraoka, M; Kulwichit, W; Raab-Traub, N; Kikutani, H. Mimicry of CD40 signals by Epstein-Barr virus LMP1 in B lymphocyte responses. *Science*, 1999, Oct 8, 286(5438), 300-3.

[13] Eliopoulos, AG; Dawson, CW; Mosialos, G; Floettmann, JE; Rowe, M; Armitage, RJ; Dawson, J; Zapata, JM; Kerr, DJ; Wakelam, MJ; Reed, JC; Kieff, E; Young, LS. CD40-induced growth inhibition in epithelial cells is mimicked by Epstein-Barr Virus-encoded LMP1: involvement of TRAF3 as a common mediator. *Oncogene*, 1996, Nov 21, 13(10), 2243-54.

[14] Li, H; Minarovits, J. Host cell-dependent expression of latent Epstein-Barr virus genomes: regulation by DNA methylation. *Adv Cancer Res.*, 2003, 89, 133-56.

[15] Waterland, RA. Epigenetic mechanisms and gastrointestinal development. *J Pediatr*, 2006, Nov, 149(5 Suppl), S137-42.

[16] Gibney, ER; Nolan, CM. Epigenetics and gene expression. *Heredity*, 2010, May 12. [Epub ahead of print]

[17] Wong, TS; Man, OY; Tsang, CM; Tsao, SW; Tsang, RK; Chan, JY; Ho, WK; Wei, WI; To, VS. MicroRNA let-7 suppresses nasopharyngeal carcinoma cells proliferation through downregulating c-Myc expression. *J Cancer Res Clin Oncol*, 2010, May 4.

[18] Wong, TS; Tang, KC; Kwong, DL; Sham, JS; Wei, WI; Kwong, YL; Yuen, AP. Differential gene methylation in undifferentiated nasopharyngeal carcinoma. *Int J Oncol*, 2003, Apr, 22(4), 869-74.

[19] Niemhom, S; Kitazawa, S; Kitazawa, R; Maeda, S; Leopairat, J. Hypermethylation of epithelial-cadherin gene promoter is associated with Epstein-Barr virus in nasopharyngeal carcinoma. *Cancer Detect Prev.*, 2008, 32(2), 127-34. 15.

[20] Seo, SY; Kim, EO; Jang, KL. Epstein-Barr virus latent membrane protein 1 suppresses the growth-inhibitory effect of retinoic acid by inhibiting retinoic acid receptor-beta2 expression via DNA methylation. *Cancer Lett*, 2008, Oct 18, 270(1), 66-76.

[21] Tsai, CN; Tsai, CL; Tse, KP; Chang, HY; Chang, YS. The Epstein-Barr virus oncogene product, latent membrane protein 1, induces the downregulation of E-cadherin gene expression via activation of DNA methyltransferases. *Proc Natl Acad Sci U S A*, 2002, Jul 23, 99(15), 10084-9.

[22] Chen, HC; Chen, GH; Chen, YH; Liao, WL; Liu, CY; Chang, KP; Chang, YS; Chen, SJ. MicroRNA deregulation and pathway alterations in nasopharyngeal carcinoma. *Br J Cancer*, 2009, Mar 24, 100(6), 1002-11.

[23] Lu, F; Weidmer, A; Liu, CG; Volinia, S; Croce, CM; Lieberman, PM. Epstein-Barr virus-induced miR-155 attenuates NF-kappaB signaling and stabilizes latent virus persistence. *J Virol*, 2008, Nov, 82(21), 10436-43.

[24] Xia, T; O'Hara, A; Araujo, I; Barreto, J; Carvalho, E; Sapucaia, JB; Ramos, JC; Luz, E; Pedroso, C; Manrique, M; Toomey, NL; Brites, C; Dittmer, DP; Harrington, WJ; Jr. EBV microRNAs in primary lymphomas and targeting of CXCL-11 by ebv-mir-BHRF1-3. *Cancer Res.*, 2008, Mar 1, 68(5), 1436-42.

[25] Lo, AK; To, KF; Lo, KW; Lung, RW; Hui, JW; Liao, G; Hayward, SD. Modulation of LMP1 protein expression by EBV-encoded microRNAs. *Proc Natl Acad Sci U S A*, 2007, Oct 9, 104(41), 16164-9.

In: Oncoproteins: Types and Detection
Editor: Jeremy R. Davis
ISBN: 978-1-61761-551-1

Chapter 6

Oncoprotein in Tropical Infection: A Story in Tropical Medicine

Viroj Wiwanitkit
Wiwanitkit House, Bangkhae, Bangkok Thailand; and Visiting Professor of Tropical Medicine, Hainan Medical College, China

Abstract

Generally, oncoprotein is mainly used in the oncological investigation and study. However, the non oncological aspect of oncoprotein is of interest. In this specific article, the author will discusses on the aspect of oncoprotein in tropical infection. This is a really forgotten aspect of oncoprotein in tropical medicine. The scenario in several tropical infections including malaria will be presented in this brief article.

Introduction

Oncoprotein is a protein coded by an oncogene. Oncoprotein is generally involved in the regulation or synthesis of proteins linked to tumorigenic cell growth [1]. Some proteins that are detectable in the cancerous condition are also sometimes - called oncoproteins and these proteins are accepted as tumor markers [1 - 2]. Generally, oncoprotein is mainly used in the oncological investigation and study. At present, advent reaches to the step that we can see the oncogene expression via probe-based imaging [3]. However, the non oncological aspect of oncoprotein is of interest. In this specific article, the author will discusses on the aspect of oncoprotein in tropical infection. This is a really forgotten aspect of oncoprotein in tropical medicine. The scenario in several tropical infections including malaria will be presented in this brief article.

Table 1 Other interesting reports on oncoprotein and malaria

Authors	Details
Dascalescu et al [10]	Dascalescu et al suggest thated the detection of FR3/JH rearrangement might contribute to the diagnosis of TSLVL in patients with malaria related chronic splenomegaly, for whom bcl-2/JH rearrangement may be a common molecular event [10].
Kumar et al [11]	Kumar et al monitored the structural integrity of mitochondria in cerebral malaria (CM) infected brain tissue by transmission EM (TEM) studies. and reported that "Our results underscore the activation of an intrinsic cell death pathway as evinced by the induction of mitochondria associated apoptotic proteins Bcl(2), Bax and cytochrome-c and further envisages the induction of p53 as a possible continuation of the post receptor signaling events associated with tumor necrosis factor induction following inflammatory responses during CM [11]."

Oncoprotein in Malaria

Malaria is a common tropical mosquito borne infectious disease that attacks millions of world population. There are several tropical endemic areas of malaria. The common features of fever with chill are described as the main clinical characteristics of malaria. It is accepted that malaria is still the present public health problem of the world. Focusing on the reports on oncoprotein in malaria, there are some interesting reports on this topic. The early hypothesis on oncoprotein in malaria was proposed in early 1990's [4 – 5]. Lalaiants et al hypothesized that "Based on similarity of valine substitution in p21 protein and hemoglobin, a relationship between certain forms of malignant transformation and malaria at molecular level is suggested [4]."

There are also a group of reports on the Duffy blood group, malaria and oncoprotein. The chemokine receptor property is proposed in general [6]. Peiper et al proposed that the Duffy antigen It is noted that the mutant of melanoma growth stimulating is the oncoprotein that has a strong relationship with this process [6]. In addition, Hesselgesser et al also reported that the mutant of melanoma growth stimulating activity did not activate neutrophils Recently, Mohammed et al additionally reported that human neutrophil lipocalin could be used as a specific marker for neutrophil activation in severe Plasmodium falciparum malaria

Oncoprotein in Dengue

Dengue is another important tropical borne mosquito infection. This tropical disease presents with a common triad of high fever, thrombocytopenia and hemoconcentration. The causative pathogen is the dengue virus. It is accepted that dengue is still the present public health problem of the world. Focusing on the reports on oncoprotein in dengue, there are some interesting reports on this topic. As a virus, the oncogenic property of dengue virus is widely investigated. The apoptosis process induced by dengue virus is reported in some papers. ApoptoM-activated signalling pathways is widely mentioned for the dengue induced apotosis [12]. Courageot et al reported that "The cellular factors that regulate cell death, such as Bcl-2 family members, can modulate the outcome of DEN virus infection in cultured cells [13]." In addition, Su et al studied on the effect of human bcl-2 and bcl-X genes and reported that "DEN infection may trigger target cells to undergo morphologically similar but biochemically distinct apoptotic pathways in a cell-specific manner [14]."

Oncoprotein in Tuberculosis

Tuberculosis is a mycobacterium infection. This is a chronic granulomatous disease. Tuberculosis can be seen at any organs of human beings but lung is the most common involved sites. In the present era, this disease can be seen in any country of the world due to the widespread of human immunodeficiency virus. It is accepted that tuberculosis is one of the most problematic infection in the present day. It is accepted that tuberculosis is still the present public health problem of the world. Focusing on the reports on oncoprotein in tuberculosis, there are some interesting reports on this topic. Similar to dengue, the oncoprotein and its correlation to apotosis is widely mentioned [17]. Important reports are summarized in Table 3.

Table 2 Other interesting reports on oncoprotein and dengue

Authors	Details
Jan et al [15]	Jan et al reported on the potential dengue virus-triggered apoptotic pathway in human neuroblastoma cells and noted that arachidonic acid, superoxide anion, and NF-kappaB were sequentially involved in the described process [15].
Lin et al [16]	Lin et al studied on endothelial cell apoptosis induced by antibodies against dengue virus nonstructural protein 1 via production of nitric oxide and reported that the expression of Bcl-2 and Bcl-x(L) decreased in both mRNA and protein levels, whereas p53 and Bax increased after anti-NS1 treatment [16].

Table 3 Interesting reports on oncoprotein, tuberculosis and apoptosis

Authors	Details
Patel et al [18]	Patel et al reported on impaired M. tuberculosis-mediated apoptosis in alveolar macrophages from HIV+ persons and proposed for the potential role of IL-10 and BCL-3 [18].
Mustafa et al [19]	Mustafa et al reported on reduced apoptosis and increased inflammatory cytokines in granulomas caused by tuberculous compared to non-tuberculous mycobacteria and further proposed on the role of MPT64 antigen in apoptosis and immune response [19].
Zhang et al [20]	Zhang et al reported that survival of virulent *Mycobacterium tuberculosis* involved preventing apoptosis induced by Bcl-2 upregulation [20].
Sly et al [21]	Sly et al reported that survival of Mycobacterium tuberculosis in host macrophages was related to resistance to apoptosis dependent upon induction of antiapoptotic Bcl-2 family member Mcl-1 [21].

References

[1] Ascione, R; Sacchi, N; Watson, DK; Fisher, RJ; Fujiwara, S; Seth, A; Papas, TS. Oncogenes: molecular probes for clinical application in malignant diseases. *Gene Amplif Anal*, 1986, 4, 253-77.

[2] Chatterjee, SK; Zetter, BR. Cancer biomarkers *Future Oncol*, 2005, Feb, 1(1), 37-50.

[3] Mukherjee, A; Wickstrom, E; Thakur, ML. Imaging oncogene expression. *Eur J Radiol*, 2009, May, 70(2), 265-73.

[4] Lalaiants, IE; Milovanova, LS. The role of "valine substitution" in oncogene functioning (a hypothesis). *Vopr Onkol.* 1991, 37(2), 206-10.

[5] Lalaiants, IE; Milovanova, LS. Valine substitution. *Biofizika*, 1990, Mar-Apr, 35(2), 231-5.

[6] Horuk, R; Chitnis, CE; Darbonne, WC; Colby, TJ; Rybicki, A; Hadley, TJ; Miller, LH. A receptor for the malarial parasite *Science*, 1993, Aug 27, 261(5125), 1182-4.

[7] Peiper, SC; Wang, ZX; Neote, K; Martin, AW; Showell, HJ; Conklyn, MJ; Ogborne, K; Hadley, TJ; Lu, ZH; Hesselgesser, J; Horuk, R. The Duffy antigen *J Exp Med*, 1995, Apr 1, 181(4), 1311-7.

[8] Hesselgesser, J; Chitnis, CE; Miller, LH; Yansura, DG; Simmons, LC; Fairbrother, WJ; Kotts, C; Wirth, C; Gillece-Castro, BL; Horuk, R. A mutant *J Biol Chem.*, 1995, May 12, 270(19), 11472-6.

[9] Mohammed, AO; Elghazali, G; Mohammed, HB; Elbashir, MI; Xu, S; Berzins, K; Venge, P. Human neutrophil lipocalin: a specific marker for neutrophil activation in severe Plasmodium falciparum malaria *Acta Trop*, 2003, Jul, 87(2), 279-85.
[10] Dascalescu, CM; Rosenzwajg, M; Bonte, H; Mir, IA; Aoudjhane, M; Smadja, NV; Lemoine, FM; Najman, A. Bcl-2 and immunoglobulin gene rearrangements in patients with malaria related chronic splenomegaly. *Leuk Lymphoma.* 2004, Oct, 45(10), 2093-7.
[11] Kumar, KA; Babu, PP. Mitochondrial anomalies are associated with the induction of intrinsic cell death proteins-Bcl(2), Bax, cytochrome-c and p53 in mice brain during experimental fatal murine cerebral malaria. *Neurosci Lett*, 2002, Sep 6, 329(3), 319-23.
[12] Catteau, A; Kalinina, O; Wagner, MC; Deubel, V; Courageot, MP; Desprès, P. Dengue virus M protein contains a proapoptotic sequence referred to as Apopto M. *J Gen Virol*, 2003, Oct, 84(Pt 10), 2781-93.
[13] Courageot, MP; Catteau, A; Desprès, P. Mechanisms of dengue virus-induced cell death. *Adv Virus Res.* 2003, 60, 157-86.
[14] Su, HL; Lin, YL; Yu, HP; Tsao, CH; Chen, LK; Liu, YT; Liao, CL. The effect of human bcl-2 and bcl-X genes *Virology*, 2001, Mar 30, 282(1), 141-53.
[15] Jan, JT; Chen, BH; Ma, SH; Liu, CI; Tsai, HP; Wu, HC; Jiang, SY; Yang, KD; Shaio, MF. Potential dengue virus-triggered apoptotic pathway in human neuroblastoma cells: arachidonic acid, superoxide anion, and NF-kappaB are sequentially involved. *J Virol.* 2000, Sep, 74(18), 8680-91.
[16] Lin, CF; Lei, HY; Shiau, AL; Liu, HS; Yeh, TM; Chen, SH; Liu, CC; Chiu, SC; Lin, YS. Endothelial cell apoptosis induced by antibodies against dengue virus nonstructural protein 1 via production of nitric oxide. *J Immunol.* 2002, Jul 15, 169(2), 657-64.
[17] Tang, SJ; Xiao, HP. Apoptosis and tuberculosis. *Zhonghua Jie He He Hu Xi Za Zhi*, 2004, Jul, 27(7), 477-9.
[18] Patel, NR; Swan, K; Li, X; Tachado, SD; Koziel, H. Impaired, M. tuberculosis *J Leukoc Biol.*, 2009, Jul, 86(1), 53-60.
[19] Mustafa, T; Wiker, HG; Mørkve, O; Sviland, L. Reduced apoptosis *Clin Exp Immunol*, 2007 Oct, 150(1), 105-13.
[20] Zhang, J; Jiang, R; Takayama, H; Tanaka, Y. Survival of virulent Mycobacterium tuberculosis *Microbiol Immunol*, 2005, 49(9), 845-52.
[21] Sly, LM; Hingley-Wilson, SM; Reiner, NE; McMaster, WR. Survival of Mycobacterium tuberculosis *J Immunol*, 2003, Jan 1, 170(1), 430-7.

In: Oncoproteins: Types and Detection
Editor: Jeremy R. Davis
ISBN: 978-1-61761-551-1

Chapter 7

False Positive in Laboratory Investigation for Oncoprotein: A Diagnostic Pitfall

Viroj Wiwanitkit
Wiwanitkit House, Bangkhae, Bangkok Thailand

Abstract

Generally, laboratory investigation for oncoprotein is a new set of laboratory investigation in oncology. The investigation plays very important roles in diagnosis and following up of cancerous patients. Similar to any investigation in laboratory medicine, the problem of laboratory investigation can be expected. The problem of false positive in laboratory investigation for oncoprotein is an interesting topic in medical oncology. Since it can lead to diagnostic pitfall, the concern on this problem in required. In this specific article, the author hereby briefly discuss on this specific topic.

Introduction

Oncoprotein is any protein coded by an oncogene and plays important role in the regulation or synthesis of proteins linked to tumorigenic cell growth [1]. Some oncoproteins are accepted and used as tumor markers [1-2]. Generally, laboratory investigation for oncoprotein is a new set of laboratory investigation in oncology [3]. The investigation plays very important roles in diagnosis and following up of cancerous patients. Similar to any investigation in laboratory medicine, the problem of laboratory investigation can be expected. Although the clinical usefulness of oncoprotein investigation is no doubtful, there are still some specific concerns on the techniques of investigation. The problem of false positive in laboratory investigation for oncoprotein is an interesting topic in medical oncology. There are some limited reports on this interesting topic on oncoprotein. Since it can lead to diagnostic pitfall, the concern on this problem in required. In this specific article, the author hereby briefly discuss on this specific topic.

What is False Positive?

In laboratory medicine, for any diagnostic test, it is needed to have the complete data on its diagnostic property. The diagnostic properties to be accessed include sensitivity, specificity, accuracy, false positive and false negative. The sensitivity and specificity are the two parameters that might be well-known to general practitioner. The sensitivity means the opportunity of have true positive result from overall derived positive result while the specificity means the opportunity to have true negative result from overall derived negative result. Focusing of false positive, it can be calculated by "false positive = 1 - sensitivity." This can be calculated for any diagnostic test.

The importance of false positive is the implication of the false diagnostic pitfall. The false positive might lead the decision of unnecessary management based on the false decision making. This means the lost of unnecessary additional investigation and drug as well as the possible danger of medication or surgery to the patients. It is the rule to get the confirmed evidence before starting therapy to get rid of or decrease the false positive as much as possible. In oncology, the problem of false positive must be seriously concerned since the excessive unnecessary procedures for the patient suspected for cancer seems to a critical thing to be controlled. In laboratory medicine, the investigation in oncology is also assigned to be an important group of investigation that takes high risk for getting poor compliance and response from the patients [4 – 5]. Sometimes, the problem can lead to sue of the practitioners [4].

Considering the rate of false positive in laboratory investigation for oncoprotein, there is no specific report. However, there are some reports on some specific oncoprotiens especially for HER-2. Ross et al noted that "There is considerable concern that both false-negative and false-positive result rates for testing for HER 2 status are unacceptably high in current clinical practice [6]." Many factors, including preanalytic conditions and slide-scoring procedures, and other variables might contribute to testing error rates [6]. In a recent study by Reiner-Concin et al on HER-2 immunohistochemistry, false positive results were found in 6.7% and 6.3% in tissue specimens and cell [7]. The normalization is the biggest concern. Gown et al reported that "Among the 6604 tumors, using a non-normalized immunohistochemistry scoring system, 267/872 (30.6%) of the immunohistochemistry 3+ cases proved to be fluorescence in situ hybridization nonamplified, whereas using the normalized scoring system only 30/562 (5.3%) of immunohistochemistry 3+ cases proved to be 'false positive' [8]." Liu et al recently reported on the effect of the change from 10% to 30% for the immunohistochemical HER2 Scoring criterion in breast cancer that false positive decreased from 12.2 % to 5.1 % [9]. Theoretically, the similar rate to other immunohistochemistry based tests for other oncoproteins might be expected. However, the rate might be totally different in other new PCR-based tests.

Table 1. Interesting reports on false positive in laboratory investigation for oncoprotein

Authors	Details
Hackwell et al [10]	Hackwell et al reported on an interesting problem when carrying out a published RT-PCR method to detect the CBFbeta/MYH11 transcripts associated with the inv(16)(p13q22) cytogenetic abnormality in acute myeloid leukaemia and this problem was shown to be due to amplification of part of the intronic MYH11 sequence [10].
Bloethner et al [11]	Bloethner et al reported on identification of ARHGEF17, DENND2D, FGFR3, and RB1 mutations in melanoma by inhibition of nonsense-mediated mRNA decay and concluded that the false positive rate was high, due to the lack of DNA mismatch repair gene defects [11].
Barrett et al [12]	Barrett et al studied on the Hercep Test for HER 2 detection [12]. Barrett et al concluded that "the majority of HercepTest 2+ results are not accompanied by gene amplification and represent "false positive" IHC in term of prognostic or therapeutic relevance [12]."
Kim et al [13]	Kim et al concluded that "an interpretation of the submicroscopic deletions of the BCR or ABL gene should not depend on ES-FISH [13]." The reason is due to a high percentage of cells with false-positive fusion signals which makes it difficult to interpret the submicroscopic ABL deletion [13].
Luthra et al [14]	Luthra et al noted for an interesting fact that although TaqMan RT-PCR assay coupled with capillary electrophoresis for quantification and identification of bcr-abl transcript type circumvent the requirement for individual fusion sequence quantitative polymerase chain reaction-based assays, this assay does not identify the specific fusion transcript [14]. This fact is useful to rule out false-positive results [14].

Interesting Reports on False Positive in Laboratory Investigation for Oncoprotein

There are some interesting reports on false positive in laboratory investigation for oncoprotein. The author hereby quotes on those interesting reports to be an idea for the reader for awareness of several forms of false positive. Of interest, there are several technical underlying concerns for the false positive.

References

[1] Ascione, R; Sacchi, N; Watson, DK; Fisher, RJ; Fujiwara, S; Seth, A; Papas, TS. Oncogenes: molecular probes for clinical application in malignant diseases. *Gene Amplif Anal*, 1986, 4, 253-77.

[2] Chatterjee, SK; Zetter, BR. Cancer biomarkers *Future Oncol*, 2005, Feb, 1(1), 37-50.

[3] Mukherjee, A; Wickstrom, E; Thakur, ML. Imaging oncogene expression. *Eur J Radiol*, 2009, May, 70(2), 265-73.

[4] Wiwanitkit, V. High-risk laboratory report: problem and prevention *J Med Assoc Thai*, 2008, Jul, 91(7), 1146-7.

[5] Wiwanitkit, V. Ethics of clinical pathologist *J Med Assoc Thai.*, 2006, Dec, 89(12), 2161-2.

[6] Ross, JS; Symmans, WF; Pusztai, L; Hortobagyi, GN. Standardizing slide-based assays in breast cancer: hormone receptors, HER2, and sentinel lymph nodes. *Clin Cancer Res.* 2007, May 15, 13(10), 2831-5.

[7] Reiner-Concin, A; Regitnig, P; Dinges, HP; Höfler, G; Lax, S; Müller-Holzner, E; Obrist, P; Rudas, M. Practice of HER-2 immunohistochemistry in breast carcinoma in Austria. *Pathol Oncol Res.* 2008, Sep, 14(3), 253-9.

[8] Gown, AM; Goldstein, LC; Barry, TS; Kussick, SJ; Kandalaft, PL; Kim, PM; Tse, CC. High concordance *Mod Pathol*, 2008, Oct, 21(10), 1271-7.

[9] Liu, YH; Xu, FP; Rao, JY; Zhuang, HG; Luo; Li, L; Luo, DL; Zhang, F; Xu, J. Justification of the change from 10% to 30% for the immunohistochemical HER2 Scoring criterion in breast cancer. *Am J Clin Pathol.* 2009, Jul, 132(1), 74-9.

[10] Hackwell, SM; Robinson, DO; Harvey, JF; Ross, FM. Identification of false-positive CBFbeta/MYH11 RT-PCR results. *Leukemia.* 1999, Oct, 13(10), 1617-9.

[11] Bloethner, S; Mould, A; Stark, M; Hayward, NK. Identification of ARHGEF17, DENND2D, FGFR3, and RB1 mutations in melanoma by inhibition of nonsense-mediated mRNA decay. *Genes Chromosomes Cancer.* 2008, Dec, 47(12), 1076-85.

[12] Barett, C; Magee, H; O'Toole, D; Daly, S; Jeffers, M. Amplification of the HER 2 gene in breast cancers testing 2+ weak positive by Hercep Test immunohistochemistry: false positive or false negative immunochemistry? *J Clin Pathol*, 2007, Jun, 60(6), 690-3.

[13] Kim, YR; Cho, HI; Yoon, SS; Park, S; Kim, BK; Lee, YK; Chun, H; Kim, HC; Lee, DS. Interpretation of submicroscopic deletions of the BCR or ABL gene should not depend on extra signal-FISH: problems in interpretation of submicroscopic deletion of the BCR or ABL gene with extra signal-FISH. *Genes Chromosomes Cancer.* 2005, May, 43(1), 37-44.

[14] Luthra, R; Sanchez-Vega, B; Medeiros, LJ. TaqMan RT-PCR assay coupled with capillary electrophoresis for quantification and identification of bcr-abl transcript type. *Mod Pathol.* 2004, Jan, 17(1), 96-103.

In: Oncoproteins: Types and Detection
Editor: Jeremy R. Davis
ISBN: 978-1-61761-551-1

Chapter 8

Oncogene Proteins: New Research E7 Oncoprotein of Human Papillomarvirus: Functions and Strategies of Inactivation for the Treatment of HPV-Associated Cancer

Maria Gabriella Donà
James Graham Brown Cancer Center, University of Louisville, KY, USA

Abstract

Human Papillomaviruses (HPVs) are involved in the etiology of at least 10% of all human cancers and are mainly implicated in the development of pre-malignant and malignant lesions of the ano-genital tract. In particular, genital High-Risk HPVs are responsible for nearly 100% of cases of cervical cancer, which represents the second most prevalent cancer in women worldwide.

The carcinogenic risk associated with HPV infection is primarily due to the activity of two viral oncoproteins, E6 and E7. Indeed, their expression is necessary for the maintenance of the malignant phenotype, and is always retained in HPV-positive cancer cells.

Although malignant transformation results from the synergistic and complementary effects of E6 and E7, the latter is reported to be the main oncoprotein of HPVs. One of the major aims of this review is to update the understanding of the role of E7 in HPV-associated carcinogenesis, focusing on the binding with its cellular targets and the effects induced by these associations. The interaction of E7 with its main target, the tumor suppressor protein retinoblastoma (pRb), will be described, with emphasis on the biological consequences and the domains involved. However, as E7 oncogenic properties are not only attributable to this association, the significance of other relevant interactions will be evaluated as well.

Due to the essential role of E7 in cervical carcinogenesis, the lack of homology with cellular proteins and the exclusive expression in cancer cells, this oncoprotein represents the main target for cervical cancer therapy. The therapeutic potential of strategies aimed at E7 inhibition for the treatment of HPV-associated lesions will be addressed. Among others, the use of small interfering RNA (siRNA), intracellular antibodies ("intrabodies"), ribozymes, anti-sense oligonucleotides and aptamers to knock out E7 at the gene/protein level will be covered. The results of these studies will be reviewed with the aim of highlighting E7 inhibition as a promising approach for the treatment of HPV-associated cancer.

Introduction

Human Papillomaviruses (HPVs) are species-specific viruses, which show strict tropism for the stratified squamous epithelium. Over 150 types have been identified so far that can infect either the skin (epithelial types) or the mucosal surface of the respiratory and ano-genital tract (mucosal types) [1]. Based on the pathology of the lesions they induce, mucosal HPVs can be distinguished as Low-Risk (LR) and High-Risk types (HR). Whereas the former group causes only benign lesions, such as genital warts (condilomata), the latter is involved in the development of pre-malignant and malignant lesions. HPV infection by HR-HPVs has been suggested to cause the majority of anal cancers, in addition to a subset of vulvar, vaginal, penile and head and neck cancers [2]. In particular, the association between HPV infection and cervical cancer (CC) has been uncontrovertibly established, biologically and epidemiologically. Almost 100% of CC cases are caused by HPV, with HPV16, 18, 31 and 45 being responsible for 80% of all cases [3]. According to a recent estimation, CC is the second most prevalent cancer in women, with 550,000 new cases/year and 270,000 deaths/year worldwide [4]. CC is the leading cause of cancer-related death for women in developing countries, where 80% of all cervical cancer cases occur.

The oncogenic potential of the HR-HPVs is mainly ascribed to the viral oncoproteins E6 and E7, which are essential for HPV-mediated carcinogenesis. Indeed, their oncogenic activity has been extensively documented in tissue culture and transgenic mouse model systems. E6 and E7 are necessary and sufficient to immortalize primary human epithelial cells [5, 6], although a fully transformed phenotype is acquired only after extended passages in culture or concomitant expression of activated oncogenes [7]. E6 and E7 continuous expression is required for the establishment and maintenance of the transformed phenotype [8, 9]. In fact, their expression is always retained in HPV-positive cancer cells [10, 11], whereas the expression of other early proteins, E2 and E4, and of the late proteins L1 and L2 (required for virus assembly) is usually lost [12]. This expression pattern is frequently related to the integration of the viral DNA into the cellular genome, which is observed in most invasive cancers but also in high-grade lesions of the cervix (HSILs) [13-15]. On the contrary, lesions induced by LR-HPVs do not contain integrated viral DNA, strongly suggesting that integration in the host genome is a key event in the acquisition of the malignant phenotype.

One of the major consequences of the integration is the loss of E2 expression. E2 has been reported to be a regulatory protein that, besides being involved in viral genome replication, can also repress expression of the viral oncogenes by displacing transcriptional activators from HPV long control region (LCR) [16]. Indeed, E2 expression suppresses E6 and E7 transcription and leads to growth arrest [9, 17, 18]. Upon integration, the E2 open reading frame (ORF) is often disrupted/deleted [19, 20]. Disruption of this regulatory gene results in loss of transcriptional repression, thus increasing expression of E6 and E7. The up-regulation of these oncoproteins confers a growth advantage to cells harboring an integrated viral genome versus cells containing episomal DNA. In fact, cervical cancer cells containing an integrated viral genome have little or no episomal DNA [15]. In addition, increased stability of E6-E7 mRNAs upon integration has also been reported [21]. Therefore, although the ultimate development of CC is a multistep process that relies on several events, the integration seems to be essential as it leads to over-expression of the viral oncogenes. Deregulated and prolonged expression of E6 and E7, associated with persistent infection, induces hyperproliferation and confers a genomic instability element that can cause the malignant conversion of the epithelial cells.

E7 in the Viral Life Cycle

Malignant transformation of HPV-positive cells is an unintended consequence of E6 and E7 activity upon a persistent viral infection. The primary function of these early proteins is not to transform the target cells, i.e. the oncogenic progression is not part of the normal viral life cycle, but to ensure the survival of the virus. Indeed, they play an essential role in the productive stage of the HPV life cycle [22, 23]. E7 expression has been demonstrated to be essential in the maintenance of viral episomes and extension of keratinocytes (KCs) life span [23]. In fact, HPV16-E7null genomes are deficient in inducing DNA synthesis and supporting viral DNA amplification [22]. Moreover, in cells containing E7-lacking HPV genomes, the expression of the main capsid protein, L1, is decreased [22].

During active infection, E7 is expressed at an early stage in order to establish a favourable environment for virus replication. At the beginning, the infection takes place at level of the dividing cells of the epithelial basal layer and the viral genome, maintained in episomal form, replicates along with the cellular DNA during the S-phase. As the basal cells proliferate, one of the daughter cells migrates toward the upper layers, undergoes differentiation and exits the cell cycle, losing its ability to divide. Therefore, these cells do not express anymore all the factors necessary for DNA synthesis. Nonetheless, HPV needs the target cells to provide these factors, since they are required for the replication of the viral genome. The infected cells of the suprabasal layers are then forced to proliferate, thus to be in a permanent replication-competent state, through the activity of E6 and E7. These viral oncoproteins profoundly deregulate the mechanisms of cell cycle control and apoptosis through the interactions established with their main targets, the tumor suppressors p53 and pRb respectively. The association with these key modulators of cell growth causes the inactivation of the pathways they control. E7 continuously stimulates the epithelial cells to support DNA synthesis by abrogating pRb function (see below), while p53-mediated pathways of DNA repair are lost due to E6 expression. Moreover, E7 delays the onset of differentiation, thus uncoupling proliferation and program of terminal differentiation, critical changes for virus survival and replication [24]. Ultimately, concomitant activity of E6 and E7 keeps HPV-positive KCs in a proliferating state as long as the virus needs the target cells to support the replication of its genome, until, at the final state of the productive cycle, the late genes are expressed and the virions are assembled.

It is worth noting that small DNA tumor viruses, such as Papillomaviruses, Adenoviruses and Polyomaviruses, have all independently evolved extremely similar mechanisms to subvert the same cellular pathways. In fact, these viruses have "developed" early proteins with remarkable functional similarity. Adenovirus (Ad) E1A and E1B, Polyomavirus Large Tumor antigen (T-Ag), HPV-E6 and E7 ultimately perform analogous tasks, in order to establish the most favourable environment for the viral replication. Some of these proteins share a large similarity in their amino acid sequence (see below) but, most interestingly, they also share the same privileged targets. HPV-E7, Ad E1A and Polyomaviruses (Simian Virus 40, JC virus and BK virus) T-Ag interact with pRb (and the other two members of the retinoblastoma family) via an LXCXE motif (see below), disrupting pRb-mediated control of cell proliferation [25-28]. On the other hand, Ad E1B, SV40 T-Ag and HPV-E6, even if structurally unrelated, show the ability of associating the tumor suppressor p53 [26, 27, 29]. The ultimate result of these interactions is the suppression of p53 function, though this is achieved through different mechanisms. In fact, E1B directly represses p53-dependent transcription by binding p53 transactivation domain [26], while T-Ag blocks p53 activity interacting with its DNA-binding domain [27]. Interestingly, p53 is stabilized (but still inactive) upon T-Ag binding, whereas in HPV-positive cells p53 is degraded as a consequence of E6-induced targeting of this tumor suppressor to the proteasome pathway [30, 31]. Recently, a novel mechanism to inactivate p53 activity has been reported for cutaenous HPVs associated with nonmelanoma skin cancer (NSSC), such as HPV38. In this case, p53 is not degraded but stabilized upon phosporylation, though it is kept functionally inactive through accumulation of its inhibitor ΔNp73 [32].

Even if the mechanisms that small DNA tumor viruses exploit to inactivate these key modulators of proliferation are not exactly the same, and this inactivation can be achieved by two viral products or simultaneously exerted by the same protein (as for T-Ag), the ultimate effect of the activity of the viral oncoproteins is the subversion of pRb and p53 pathways. This strengthens the notion that the concomitant inactivation of these tumor suppressors has a primary importance for the efficient replication of these viruses. In fact, inactivation of pRb is required to promote the proliferation of the infected cells in order to render available the cellular factors necessary for the replication of the viral genome. On the other hand, suppression of p53 pathway guarantees the survival of the infected cells, since it abrogates the cellular responses, such as apoptosis, that normally protect the cells in case of abnormal proliferation.

E7 as an Immortalizing and Transforming Protein

Several *in vitro* studies on E7 proteins of HR-HPVs have extensively demonstrated its transforming activity, in contrast with E7 proteins of LR-HPVs that do not show any oncogenic potential [5, 6, 33]. E7 is necessary and sufficient to transform established murine fibroblast cell lines [34]. The product of HPV16 and 18 shows the ability of transforming murine primary cells in cooperation with the activated *ras* oncogene [35, 36]. Continuous expression of E7 is required for the growth of the transformed cells and their tumorigenicity in nude mice. In addition, E7 is sufficient to extend the life span of human primary keratinocytes, although their immortalization also requires E6 or the entire HPV genome [5]. HPV16-E7 also immortalizes human foreskin keratinocytes (HFKs) when co-expressed either with Myc or with human telomerase (hTERT) [37]. As in the case of E6/E7-induced immortalization, co-expression of E7 and Myc induces telomerase activation [37]. In fact, HPV16-E6 (but not HPV16-E7) activates hTERT promoter through a c-myc-dependent mechanism [38, 39]. However, some studies report that E7 expression is sufficient to immortalize human epithelial cells [40, 41].

E7 oncogenic potential has been confirmed by additional studies on transgenic mice. In fact, mice expressing the HPV16-E7 oncoprotein in the basal layer of the squamous epithelium show hyperplasia, increased DNA synthesis in basal and suprabasal layers, loss of DNA damage response and lack of or delayed differentiation [42, 43]. E7 alone shows the capacity to induce both high-grade dysplasia and invasive cervical cancer in transgenic mice treated with estrogens [44].

E7 can abrogate several signals of growth arrest, inducing escape of quiescent cells from G0 or premature entrance of proliferating cells in S-phase [45-47]. Several studies have shown that E7 can overcome growth arrest signals provided by a number of cytokines with cytostatic effect on epithelial cells. For instance, resistance to TNFα, a cytokine involved in regulation of proliferation and differentiation of keratinocytes, has been reported [48]. Normal KCs respond to TNFα with arrest in G1, as a consequence of p53-independent $p21^{CIP1/WAF1}$ induction. On the contrary, E7-positive KCs continue to proliferate upon TNFα treatment. E7 ability to overcome the cytostatic signal provided by TNFα seems to be correlated with "pocket proteins" inactivation [49] (see below). E7 protein of HR-HPVs can also abolish growth suppression induced by TGFβ [50].

E7 Functional Domains

The E7 oncogene encodes a small acidic phosphoprotein of approximately 100 amino acid residues, in particular 98 for the HPV16 product and 105 for the HPV18 product.

As already mentioned, E7 shows similarity with two other oncoproteins of DNA tumor viruses: Ad E1A and SV40 T-Ag [36]. Based on this similarity, E7 molecule can be divided in three functional regions, namely "conserved region 1" (CR1), CR2 and CR3, spanning the amino acids 1-15, 16-37 and 38-98 respectively for the product of HPV16 (Figure 1). The CR2 contains the high affinity pRb binding site (see below) and a consensus phosphorylation site for the casein kinase II (CKII) [51]. Additionally, the presence of another phosphorylation site has been reported at the C-terminus, though this represents the target of a not yet identified kinase [52].

Although the structural and functional homology between E7 and E1A is restricted to the N-terminus moiety, both oncoproteins display the ability to chelate zinc through the C-terminus. In E7 this property is due to the presence of two Cys-X-X-Cys motifs, usually spaced by 29 amino acids. The structure and length of E7 zinc-finger is unusual and unique and is not shared by any cellular protein with zinc finger DNA-binding domains [53, 54]. Changes in the length of the spacer between the motifs [55] as well as mutations in one or both cysteines in one or both Cys-X-X-Cys motifs [23, 53, 56] have been reported to affect E7 intracellular stability. Furthermore, the zinc-binding domain seems to be responsible for

E7 ability to dimerize/oligomerize [53, 57]. Most interestingly, it is involved in a large number of E7 biological activities.

Indeed, an intact zinc-finger is required to immortalize human keratinocytes [58] and mutations in this domain impair E7 ability of transforming rat primary embryo fibroblasts in cooperation with activated *ras* oncogene [53].

In recent studies E7 has been described as an "intrinsically disordered" or "natively unfolded" protein. In fact, data collected by circular dichroism have suggested the presence of unfolded regions, together with some ordered secondary structure [59, 60]. More than 50% of the residues of E7 proteins from HR-HPVs have been identified as disordered [61]. In particular, recent results of the solution structure of E7 protein from HR-HPV45 have shown that the N-terminus is completely unfolded, while the C-terminus is well-structured [54]. This ordered region contains the zinc-binding domain [54, 61].

E7 has been reported to have a half-life between 1 and 2 hours in HPV16-positive CaSki and SiHa cells [62, 63]. This relative instability likely explains the relatively low steady-state level of the oncoprotein in cervical cancer cells. Indeed, it has been evidenced that E7 can be rapidly degraded by the proteasome pathway [64, 65]. However, this degradation does not involve classical ubiquitin conjugation to internal lysine residues, but depends on conjugation at the N-terminus [64].

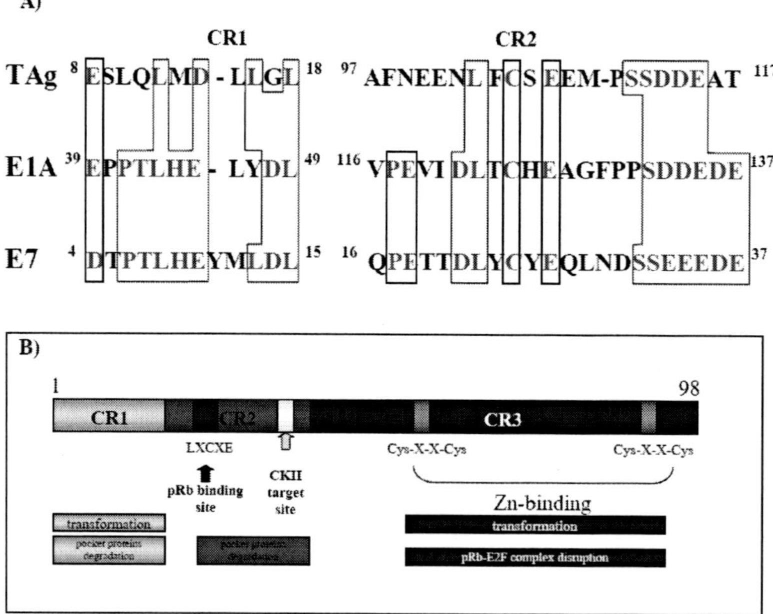

Figure 1. Sequence similarity among SV40 TAg, Ad E1A and HPV16-E7 and schematic representation of HPV-E7 domains. A) The primary sequence of Adenovirus 5 E1A oncoprotein conserved regions (CR) 1 and 2 is compared to SV40-TAg and HPV16-E7 homologous regions. The conserved amino acids are indicated in red. The position of the residues starting form the N-terminus is indicated. B) The three structural/functional domains of HPV-E7 are schematically illustrated. The LXCXE motif, representing pRb high-affinity binding site, is indicated. Moreover, the localization of the target site for the casein kinase II (CKII) is reported. The position of the Cys-X-X-Cys motifs involved in the formation of E7 Zn-finger is also shown.

Studies on E7 localization have been greatly impaired by the lack of sensitive antibodies and most of the available data have been obtained in transfected cells over-expressing tagged forms of the oncoprotein. However, some observations have also been accomplished in HPV-positive cervical carcinoma cell lines. The majority of the studies have indicated E7 as a mainly nuclear protein [66-68]. Interestingly, a preferential staining of nucleoli has been evidenced in HPV16-positive cells [66]. Localization in nucleolus has also been reported when E7 has been expressed in yeast cells, suggesting that a conserved mechanism is involved in its nucleolar transportation [66]. Since a canonical nuclear localization signal has not been identified, it is possible that E7 accumulation in the nucleus results from its association with cellular proteins that serve as a shuttle. However, since the fission yeast appears to lack the "pocket proteins", their role in serving as a shuttle for E7 has been excluded [66]. The mechanism of E7 nuclear import has not been clarified yet.

Since several cytoplasmic targets of E7 have also been described, E7 presence in the cytoplasm cannot be ruled out. Indeed, E7 cytoplasmic localization has been reported by some studies [62, 66].

E7 and the Retinoblastoma Family of "Pocket Proteins"

Retinoblastoma (pRb)

E7 is a multifunctional protein that can modulate the activity and/or stability of more than 20 cellular proteins, including cell cycle regulators, transcription factors and metabolic enzymes (Table 1). Through the interactions established with its numerous targets, E7 influences G1-S check-point transition, cellular differentiation, apoptosis, host immune response and cell metabolism.

Numerous observations have suggested that HPV-induced malignant transformation is intimately linked to E7 interference with cell cycle control pathways. Indeed, several cellular proteins involved in the control of cell proliferation represent the main targets of E7. The most relevant and well-characterized interactions are those established with the members of the retinoblastoma family of "pocket proteins": pRb [28], p107 and p130 [69]. The "pocket proteins" modulate the progression from G0 to G1 and the transition between G1 and S-phase, besides being implicated in regulation of differentiation, apoptosis and DNA damage response [70].

The "pocket proteins" function in cell cycle control is mainly accomplished through the interaction with the E2F family of transcription factors, which includes at least eight closely related proteins (reviewed in [71]). pRb is mainly found in complexes with E2F1 through E2F3, which are typically associated with active promoters in S-phase. These transcription factors control the expression of genes necessary for nucleotide synthesis (dihydrofolate reductase and thymidine kinase), DNA replication (DNA polimerase α) and cell cycle progression (cyclin A and E, c-myc; [72]).

pRb represses these proliferation-associated genes by directly blocking E2F-mediated transactivation. Additionally, pRb actively represses E2F-responsive promoters recruiting chromatin remodelling enzymes, such as the histone deacetylases (HDACs) [73].

Table 1.

Target	Functional Effect	References
pRb	Disruption of E2F complexes Degradation	[28, 75, 76]
p107	Disruption of E2F complexes Degradation	[69, 77]
p130	Disruption of E2F complexes Degradation	[69, 78]
$p21^{WAF-1}$	Inactivation of cdk inhibitory activity	[79, 80]
$p27^{KIP1}$	Inactivation of cdk inhibitory activity	[81]
AP-1	Activation of transcription	[82]
TBP	?	[83]
Mi2β (HDACs)	Modulation of transcription	[84]
M2-PK	?	[85]
α-glucosidase	?	[86]
S4 (26S proteasome)	?	[87]
IRF-1	Inhibition of IFN pathways	[88, 89]
TAF110	?	[90]
p48 (ISGF3)	Inhibition of IFNα stimulation	[91]
F-actin	?	[92]
MPP2	Activation of MPP2 transcriptional activity	[93]
hTid	?	[94]

pRb growth suppression function is regulated by a series of complex interactions, which also involve the tumor suppressor gene p53.

In fact, besides regulating pRb expression, p53 also modulates its activity by inducing the transcription of the inhibitor of cyclin-dependent kinases $p21^{CIP1/WAF1}$ that maintains pRb in the active form (see below) [74].

In turn, pRb modulates p53 activation since E2F also controls the expression of proteins that indirectly induce p53 pathway, such as the alternative reading frame (ARF) and ataxia-telangiectasia mutated (ATM) proteins. These proteins inhibit the activity of MDM2, which normally promotes p53 degradation. Importantly, p53 also represents a target of HPV. As already mentioned, the viral oncoprotein E6 suppresses p53 pathway by inducing degradation of this tumor suppressor via association with the cellular ubiquitin-protein ligase E6-associated protein (E6AP) [30, 31, 95].

The biological activity of retinoblastoma is regulated in response to growth factors through signalling pathways that lead to pRb phosphorylation, which represents the primary mechanism of control of pRb function in cell cycle regulation. In fact, pRb phosphorylation state determines pRb binding to its target regulatory proteins. Retinoblastoma is phosphorylated at several serine residues by G1 cyclin-dependent kinases (cdks) associated with their respective cyclins (cyc). In particular, cycD/cdk4-6 complexes first and cycE/cdk2 complexes later are mostly responsible for pRb phosphorylation, which begins at the G1/S boundary [96]. Dephosporylation by a specific phosphatase occurs at late M phase. In the hypophosphorylated form pRb binds E2F forming a transcriptional repressor complex; therefore the expression of E2F-responsive genes is inhibited. In late G1 phase, retinoblastoma phosphorylation by cdks determines the release of the E2F factor, now converted into a transcriptional activator.

A considerable amount of evidence strengthens the notion that association of E7 with pRb represents an essential event in E7-induced transformation, as the binding affinity of E7 for this target correlates with the transforming potential of the viral product [97].

E7 preferentially binds the hypophosphorylated form of pRb [69, 98], preventing its association with E2F and inducing E2F release from already established complexes [56]. In both cases, an increase in the level of free E2F is promoted, resulting in progression in S-phase (Figure 2).

Considering the new insights in E2F activity, it appears that induction of cell proliferation is not (or not only) a result of E2F-mediated transcription activation but also depends on E2F-mediated de-repression of its target genes [71].Aberrant expression of proliferation-associated genes, such as cycA, which is required during S-phase, and cycE, necessary for G1-S transition, has been detected in E7-positive cells, with concomitant activation of cdk2 [46, 99, 100]. Significantly, constitutive expression of these genes and simultaneous increase in associated kinase activities can be induced by E7 in the absence of external growth factors [46].

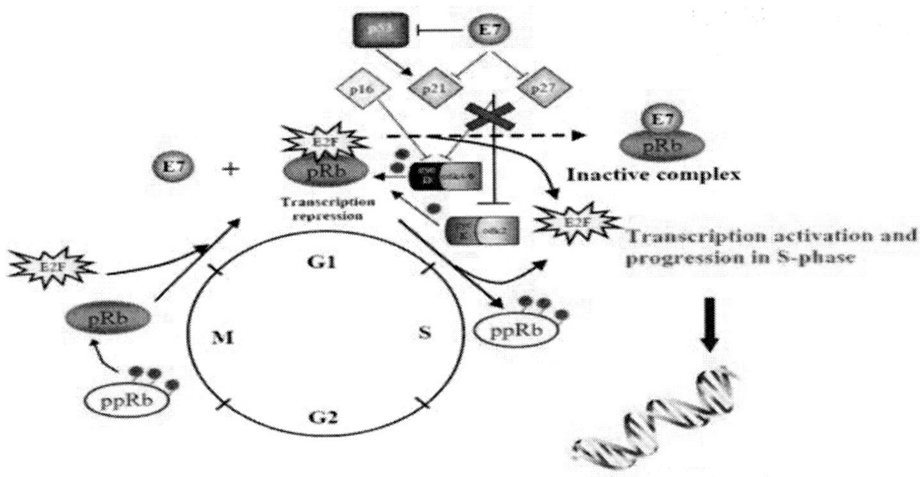

Figure 2. Schematic representation of E7-pRb interaction. In normal cells, hyposphorylated pRb binds the E2F transcription factor forming a repressor complex and repressing the expression of genes necessary for cell cycle progression. pRb phosphorylation by cyc/cdks complexes at the G1/S boundary induces E2F release and transcription of E2F-responsive genes, such as timidine kinase (TK), dihydrofalate reductase (DHFR) and DNA polymerase α (DNApolα). In HPV-positive cells, E7 binds the hyphosphorylated form of pRb and promotes E2F release, thus S-phase entry. In addition, E7 blocks the activity of the cdk inhibitors p21 and p27, which normally suppress cyc/cdks complexes.

It is worth noting that the ability of E7 to transactivate E2F-responsive genes relies on complex and concomitant subversions of several regulatory pathways, through mechanisms that are not only linked to direct impairment of pRb activity. In fact, it has been shown that E7 can activate E2F1-driven transcription even in absence of pRb [101]. It has been demonstrated that E7 can interact with E2F1 both *in vitro* and *in vivo* and that, most interestingly, can form complexes with E2F1-DNA. These data suggest that, besides the interaction with pRb, the direct association with E2F might have a functional role in E7-induced transformation by promoting the expression of proliferation-associated genes.

Importantly, E7-mediated induction of E2F-responsive genes encoding cycA and cycE allows functional bypass of p16ink4a interference with cellular proliferation. This protein is a cdk-inhibitor (CKI) that suppresses the activity of cycD/cdk4-6 complexes, responsible for pRb phosphorylation in mid-late G1. Therefore, p16ink4a over-expression normally results in cell cycle arrest as it leads to accumulation of hypophosphorylated pRb and sequestration of E2F [102]. However, in the presence of E7 p16ink4a-mediated G1 arrest is bypassed, since pRb is inactivated/degraded as a result of the oncoprotein activity [103, 104]. Indeed, E7-mediated pRb degradation is required for the efficient suppression of G1 block imposed by p16ink4a [104].

Since E7-mediated impairment of pRb function causes substantial up-regulation of p16ink4a, E7-positive cells show aberrant p16ink4a accumulation, as frequently observed in HPV-associated neoplastic lesions and CC [105]. Interestingly, cervical biopsies from high-grade lesions of the cervix show strong staining for the serine/threonine kinase AKT as well [106]. Enhanced activity of AKT in E7-positive keratinocytes has been reported [106, 107] and also in this case a correlation with pRb impairment has been evidenced [106]. The biological relevance of promoting AKT function in cervical carcinogenesis has not been completely clarified. However, some observations have suggested its involvement in E7-induced HFKs migration upon $p27^{KIP1}$ mislocalization (see below) [107]. Moreover, the phosphoinositide 3 kinase (PI3K)-protein kinase B (PKB)/AKT pathway mediates resistance to anoikis, an apoptotic response that is activated in epithelial cells upon disruption of their interaction with the matrix [108]. Importantly, it has been shown that HPV16 E6/E7 can confer HaCaT cells resistance to anoikis and that this effect is enhanced when the viral oncogenes are co-expressed with a constitutively activated form of PKB/AKT [109].

E7 association with pRb is mediated by an LXCXE motif, which represents the canonical binding site for "pocket proteins". This motif is shared by many cellular proteins that interact with retinoblastoma, such as D-type cyclins, HDAC-1 and HDAC-2, besides the other viral oncoproteins able to functionally inactivate pRb, such as Ad5 E1A and SV40 T-Ag [25, 69, 110, 111]. The LXCXE motif maps in the CR2 region, therefore it is included in the unstructured and highly flexible N-terminal domain of E7 [54, 61]. Hence, this position might represent a way to keep the pRb-binding domain always accessible to pRb. The LXCXE binding site on pRb is localized in the pRb "pocket" domain, which is necessary, although not sufficient, for pRb-mediated growth inhibition [112, 113]. This "pocket" region has been co-crystallised with a nine-residue HPV16-E7 LXCXE peptide, which binds in an extended conformation to a hydrophobic groove included in the pRb-B box [114]. In particular, HPV16-E7 Leu 22, Cys 24, Glu 26 and Leu 28 make direct contact with highly conserved amino acid residues of pRb, evidencing that residues surrounding the LXCXE motif are also important for establishing a tight interaction.

The LXCXE motif is sufficient for pRb binding [115]. Nonetheless, several data suggest that other regions of the oncoprotein are involved in pRb association. A peptide containing the LXCXE sequence is 100-fold less efficient than the full-length E7 protein in inhibiting pRb binding in a competitive binding inhibition assay [115]. Moreover, it has been shown that E2F, which does not contain an LXCXE motif, and E7-LXCXE binding sites on pRb pocket domain do not overlap but map at two different sites, which are 30 Å apart [116]. Therefore, the LXCXE sequence alone cannot explain either the association or the functional inactivation of pRb. The C-terminus of E7 has been reported to contain a low-affinity binding site for pRb C-terminus [115, 117]. In addition, it seems that this domain harbors binding sites for the "marked box" region of E2F as well [117]. Indeed, the CR3 region of E7 is essential in inducing E2F displacement, whereas the LXCXE motif alone is not capable of promoting the release of transactivation-competent E2F [115, 117], suggesting the model depicted in Figure 3 for E7-mediated inactivation of pRb.

Besides the direct interaction and inactivation of pRb, there is evidence that E7 induces degradation of this tumor suppressor both *in vitro* [75] and *in vivo* [43, 77]. Most interestingly, retinoblastoma degradation seems to be necessary for efficient inactivation of pRb function by E7. Indeed, an E7 mutant able to bind but not to degrade pRb is partially defective in blocking pRb activity compared to the wild-type protein [77]. Notably, E7 capacity to induce pRb degradation strongly correlates with its transforming activity [118]. In fact, HPV1a-E7 that can bind pRb with almost the same efficiency of HR-HPV E7 but does not induce pRb degradation [77] cannot transform primary cells [119, 120]. Interestingly, E7 ability of promoting pRb degradation has been shown to correlate also with the E7 capacity of circumventing p16ink4a-induced cell cycle arrest [104]. Therefore, further dysregulation of the pRb pathway by E7-induced degradation strongly contributes to E7 transforming activity, as it may enhance the expression of E2F-responsive genes required for G1/S transition and DNA synthesis. Besides the impairment of pRb activity by direct binding, the ability to promote its degradation might represent a more efficient way of neutralizing the tumor suppressor function if E7 is not present in the cell at a sufficient stoichiometric level to sequester enough pRb.

Though the exact mechanism of E7-mediated pRb degradation has not been clarified yet, the ubiquitin-proteasome pathway seems to be involved [75, 77]. Although HPV16-E7 has been shown to recruit the proteasome 26S and E7 interaction with S4 subunit of the proteasome has been demonstrated [87], S4 does not appear to be responsible for pRb targeting to the proteasome pathway. In fact, an S4 binding deficient mutant of HPV16-E7 can still induce pRb degradation [77]. A recent study has reported that the cullin 2 ubiquitin ligase might be involved in HPV16E7-mediated pRb degradation [76]. Indeed, pRb/E7 complexes have been shown to contain cullin 2. In addition, silencing of this ubiquitin ligase determines an increase in pRb steady-state level.

E7 ability of inducing pRb degradation is impaired by mutations in CR1 and CKII phosphorylation site, suggesting that pRb binding, which occurs through CR2 and CR3 domains, is not sufficient for its degradation [77]. However, the amino acids located immediately after the LXCXE motif also seem to be essential for pRb destabilization [104].

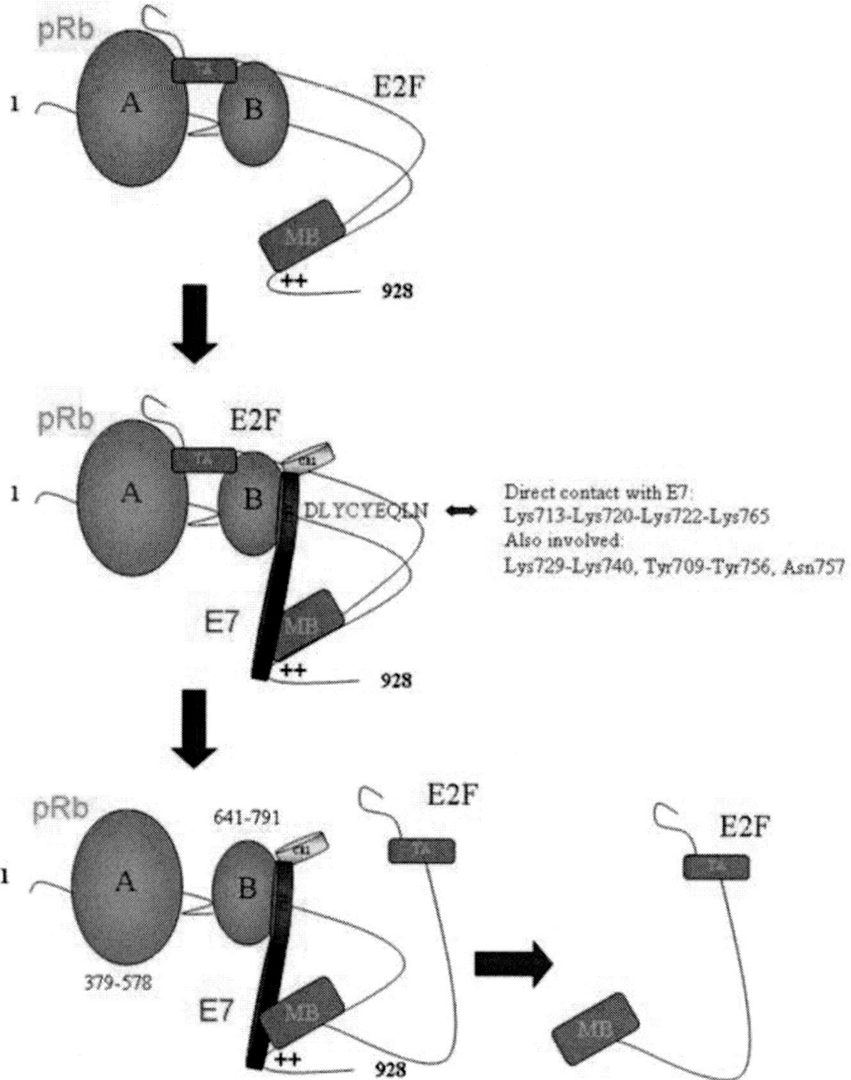

Figure 3. Schematic representation of E7-induced displacement of E2F from pRb. E2F transactivation domain (TA) binds at the interface between pRb A and B boxes (represented by the green ovals), while E2F "marked box" (MB) contacts a positively-charged region in pRb C-terminal (shown as ++). HPV-E7 CR2 domain interacts with the B box via an LXCXE domain (indicated in red). pRb amino acid residues involved in the interaction are illustrated. E7 C-terminus makes contacts both with pRb C-terminus and E2F-MB, displacing E2F probably through an allosteric mechanism. Modified from Liu *et al*, 2006 [117].

p107 and p130

Several reports have shown that E7 targets two other factors that are functionally and structurally related to the retinoblastoma protein: p107 (retinoblastoma-like 1, RBL1) and p130 (RBL2) [69, 121]. The data available suggest importance of these interactions, even though they have not yet been well characterized. In comparison with pRb, p107 and p130 have not been conclusively linked to tumor suppressor activity and their contribution to E7 oncogenic potential is still debated.

These "pocket proteins" are differentially expressed during the cell cycle: p107 is mainly synthesized during the S-phase, while p130 level is abundant in G0, therefore in differentiated and quiescent cells [122]. These proteins do not associate the E2F1-3 members of the E2F family, which are exclusive targets of pRb. Instead, they form complexes with E2F4-5. These factors, opposed to E2F1-3, act as repressors and have a reduced ability to induce S-phase entry [72].

E7 association with p107 has been shown to occur independently of pRb binding and might contribute to E7-induced immortalization. In fact, HPV16-E7 mutants defective in pRb but not p107 binding are impaired in transformation but retain the ability of immortalizing human cells [58, 121]. However, in contrast with what has been evidenced for pRb, E7 is not able to induce E2F dissociation from p107 in S-phase [123, 124], suggesting that in this case an alternative mechanism is exploited in E2F activity regulation. On the other hand, E7 has been shown to disrupt E2F-p107 complexes in G1-phase [125], although the biological significance of this different behaviour is still unclear.

E7 displays the ability of interacting with p130 as well [69]. This "pocket protein" appears to be implicated in regulation of cellular differentiation and especially in the maintenance of terminal differentiated state [126]. Indeed, differentiated cells characteristically show accumulation of p130/E2F4 complexes. Therefore, it is likely that E7-mediated inhibition of differentiation is promoted by association with p130 and alteration of its function. In fact, the ability of E7 to bind and degrade p130, besides the other "pocket proteins", correlates with its capacity of disrupting keratinocyte differentiation [24]. Hence, this activity may be more relevant in the virus life cycle than in HPV-mediated carcinogenesis.

Although some experiments failed to prove that p107 and p130 levels are affected by E7 [127], other studies have evidenced their degradation consequent to E7 expression, as observed for pRb [77, 78]. Destabilization of these proteins might further contribute to E2F-dependent transcription. Complexes of p130, HDACs and E2F4-5 are prominent on E2F-promoters in quiescent and differentiating cells. Thus, E7-induced degradation of p130 might cause the replacement of these repressor complexes with the activators E2F1-3 and induce transcription activation. However, the correlation between the ability of promoting p107 and p130 degradation and E7-induced transformation is uncertain. Since E7 of LR-HPV6 also shows the capacity to destabilize p130, this activity might be a common feature of the Papillomaviruses, necessary to uncouple proliferation and differentiation during the viral life cycle [24, 78].

E7-mediated inactivation of all the "pocket proteins" suggests that this multiple knock out might be required in the context of the HPV life cycle, for several reasons. Impairment of all pRb family members may be necessary to achieve a complete deregulation of the cell cycle, since it may allow bypassing the compensation of pRb function loss by p107 and p130. In fact, the effects of inappropriate release of activators E2Fs upon E7-induced pRb loss could be compensated by their association with the other two "pocket proteins", i.e. these proteins could substitute for pRb in the regulation of E2F1-3. Substantially, in absence of functional pRb, p107 and p130 might act as tumor suppressors (though in pRb-deficient tissue/cells this circumstance depends on the availability of the other two "pocket proteins"). Additionally, the functional inactivation of all the three members of the retinoblastoma family might indicate that HPV needs to deregulate different processes, since the activities of the "pocket proteins" are only partially redundant. For instance, pRb is more involved in proliferation and early differentiation, while p130, as already mentioned, regulates later stages of differentiation. Therefore, pRb inactivation might be relevant early in the life cycle, whereas, once the cell cycle is deregulated and the early differentiation is impaired, disruption of p130 activity may become prominent to disturb terminal differentiation [24].

Other E7 Cellular Targets Potentially Contributing to E7 Oncogenic Activity

Although pRb is described as the main target of E7, association with this tumor suppressor protein alone is not sufficient to fully explain E7 oncogenic activity. In fact, mutations in the high-affinity binding site do not abrogate E7 ability to immortalize primary keratinocytes [121], whereas mutations in HPV16-E7 regions other than the pRb-binding site, such as the CR1, the CKII target site and the C-terminus, can impair E7 immortalizing and transforming properties [128]. Moreover, some non-transforming mutants of E7 show the ability of binding pRb and transactivating E2F-responsive promoters [58, 129, 130]. Studies conducted in transgenic mice with somatic *Rb* deletion have shown that E7 can still induce hyperplasia and displasia in pRb null tissues, suggesting that pRb-independent activities are also responsible for E7 oncogenic potential [43].

Association of E7 with numerous other cellular proteins has been reported and novel targets are continuously described, although the biological relevance of these interactions is often unclear.

Cyclins and Cyclin-Dependent Kinases (Cyc/Cdks)

Interaction of E7 with cyc/cdks complexes has been observed; in particular, E7 association with cycA/cdk2 [131, 132] as well as cycE/cdk2 [132] has been reported. It is conceivable that the "pocket proteins" operate as mediators in such interactions. Indeed, E7 binding to cycE does not appear to be direct but mediated by p107 [132]. However, other results suggest a direct interaction, as for the cycA/cdk2 complex [133]. HPV16-E7 shows the specific ability to promote the function of cdk2 complexed either with cycA or cycE. Although enhanced cdk2 activity might result from inhibition of $p21^{CIP1/WAF1}$ and $p27^{KIP1}$ (see below), at least a study has reported that the activation of cdk2 directly depends on its interaction with E7 and does not involve deregulation of the cdk inhibitors [133]. The ability of enhancing cdk2 activity seems to be a conserved feature among HPVs, as it has been demonstrated for LR-HPV E7 proteins as well [133]; therefore, whether the binding of E7 to this kinase has a functional role in E7-induced transformation remains uncertain. Induction of cdk2 function could be important to promote S-phase entry during the viral life cycle, since the histone H1 kinase activity is specifically enhanced and histone H1 phosphorylation models chromatin structure in preparation of DNA synthesis [133]. Additionally, since cdk2 is essentially involved in pRb phosphorylation, induction of its activity may have the significance of further enhancing pRb inactivation.

Finally, E7 might deregulate the activity of the cyc/cdks complexes modifying the time of activation during the cell cycle or the spectrum of targets phosphorylated by the kinases.

Cyclin-Dependent Kinase-Inhibitors (CKIs)

Modulation of cyc/cdks activity and abrogation of cell cycle inhibition occur also through E7 interaction with several other proteins involved in control of cell cycle progression. For instance, the function of the CIP/KIP family of cdk inhibitors (CKIs) $p21^{CIP1/WAF1}$ and $p27^{KIP1}$ has been reported to be blocked by E7 [79-81]. The interaction with both these CKIs has been ascribed to E7 C-terminus.

Inhibition of $p21^{CIP1/WAF1}$ ability of blocking cyc/cdk function, as well as of suppressing proliferating cell nuclear antigen (PCNA)-dependent DNA replication, has been demonstrated as a result of a direct interaction between the CKI C-terminus and HPV16-E7 [79]. Perturbation of $p21^{CIP1/WAF1}$ activity by E7 may have many relevant consequences due to the involvement of this CKI in numerous processes, including response to DNA-damage, senescence and differentiation. For instance, $p21^{CIP1/WAF1}$ is implicated in cell cycle withdrawal during keratinocyte differentiation; therefore, the abolishment of its function may have a role in inducing DNA synthesis in differentiated epithelial cells [79, 134]. Since HPV needs to uncouple cellular differentiation and proliferation, as already mentioned, deregulation of this CKI can be particularly relevant for the virus life cycle. Normally, as the level of $p21^{CIP1/WAF1}$ increases during the differentiation process, cdk2 is inactivated and growth arrest is induced. However, in E7-positive cells this CKI remains inactive, thus, cdk2 remains active and $p21^{CIP1/WAF1}$-mediated growth arrest is bypassed [79, 134]. Interestingly, these effects have been observed in spite of the elevated levels of $p21^{CIP1/WAF1}$, which are due to its stabilization [135, 136]. $p21^{CIP1/WAF1}$ inactivation seems to be required together with pRb deregulation to subvert cell cycle control in human epithelial cells [137]. Significantly, suppression of this CKI represents another mechanism to abolish pRb function, since $p21^{CIP1/WAF1}$ inhibits the activity of cdk2, which is responsible for pRb phosphorylation and inactivation (see above). Substantially, the repression of CKIs further contributes to guarantee that pRb is maintained in the inactive form and that pRb-regulated cellular processes are completely impaired. Moreover, direct inactivation of $p21^{CIP1/WAF1}$ also allows bypassing p53-mediated antiproliferative pathways that normally lead to $p21^{CIP1/WAF1}$ transcriptional activation, since this CKI represents a transcriptional target of p53. Obviously, since HPV-positive cells express also E6, its ability to promote p53 degradation fundamentally contributes to block p53-mediated responses as well.

Finally, since $p21^{CIP1/WAF1}$ is critically involved in G1-arrest upon DNA damage, $p21^{CIP1/WAF1}$ blockage might promote genetic instability during HPV-associated cancer progression.

E7 interaction with $p27^{KIP1}$ has also been reported, both *in vitro* and *in vivo* [81]. As for $p21^{CIP1/WAF1}$, E7 has been shown to inhibit $p27^{KIP1}$ activity. This ability appears to be responsible for E7 capacity of overcoming growth arrest in the presence of anti-proliferative signals. For instance, $p27^{KIP1}$ inhibition seems to be implicated in E7 ability of abrogating keratinocytes growth arrest in response to TGFβ [50]. In addition, a recent study has demonstrated that HPV16-E7 can cause $p27^{KIP1}$ mislocalization in HFKs, with cytoplasmic retention of this CKI [107]. Importantly, prominent cytoplasmic staining of $p27^{KIP1}$ has been shown in cervical cancer biopsies as well [107]. It has been postulated that E7-induced $p27^{KIP1}$ mislocalization might be an important determinant in HPV16-positive tumors invasiveness [107]. In fact, cytoplasmic $p27^{KIP1}$ appears to be a positive modulator of migration and tumor invasion [138-140]. In agreement with these data, E7-positive cells with cytoplasmic $p27^{KIP1}$ show increased migration capability. E7-induced cytoplasmic retention of $p27^{KIP1}$ is achieved through subversion of PI3K/AKT pathway, which seems to be involved in $p27^{KIP1}$ localization in primary KCs, since the use of specific inhibitors for these kinases determines increase of the CKI nuclear level [107].

E7 ability to subvert $p27^{KIP1}$ function and localization may therefore account for important events in the carcinogenic process, not only at level of tumor development but also tumor invasiveness.

Cdc25A

A further disruption of cell cycle control by E7 is achieved through modulation of Cdc25 proteins, which are tyrosine phosphatases involved in cyc/cdks activation upon removal of inhibitory phosphorylation [141]. In particular, Cdc25A controls the activity of cycA and cycE-dependent kinases, dephosphorylating and activating cycE/cdk2 and cycA/cdk2 complexes. Therefore, it largely contributes to progression in S-phase. HPV16-E7 has been reported to transactivate cdc25A promoter, whereas this activity has not been detected for the product of the LR-HPV11 [142]. Several data have suggested that E7 capacity of bypassing growth arrest signals is linked to the induction of cdc25A expression. For instance, a strict correlation has been evidenced between E7-induced Cdc25A overexpression and E7 ability to promote proliferation in conditions of serum starvation, normally associated with repression of cdc25A transcription [47]. Moreover, as Cdc25A level remains high during differentiation of E7-positive human KCs, E7-driven modulation of Cdc25A may be involved in E7-induced delay of primary keratinocyte differentiation [47]. It has been shown that increased expression of Cdc25A requires intact pRb and HDAC1-binding domains, suggesting that cdc25 modulation occurs in a HDAC-dependent manner through disruption/derepression of pRb/E2F/HDAC-1 complexes (see below).

Importantly, Cdc25 phosphatases have also a significant role in regulating G2/M transition upon DNA damage. Cdc25A is normally degraded or inactivated in response to DNA damaging agents because of the activation of the ATM and ATM-related (ATR) pathways [141]. Therefore, deregulation of this phosphatase may contribute to perturb cell response to DNA injury, rendering the cell unable to stop cell cycle progression either to repair DNA or to initiate apoptosis. Ultimately, defects in Cdc25A activity may result in genomic instability. All these data strongly support the idea that Cdc25A is an important mediator in E7-induced proliferation and tumorigenesis.

Histone Deacetylases (HDACs)

The association of HR-HPVs E7 proteins with the histone deacetylases 1 and 2 (HDAC-1 and HDAC-2) has been evidenced both *in vitro* and *in vivo* [23, 84]. HDACs are three classes of proteins that repress gene expression by removing acetyl groups from the N-terminal tails of the histones. Histone deacetylation leads to nucleosome condensation, which in turn blocks the access of transcription factors to the DNA. Several studies have demonstrated pRb's ability to recruit HDAC to E2F-regulated genes and an HDAC requirement for pRb to repress several cellular genes [143, 144]. However, if HDAC recruitment is strictly necessary to mediate pRb inhibition of E2F-responsive transcription is still controversial. The histone deacetylases might be involved in counteracting the activity of the histone acetyltransferases (HATs) p300/CBP and p/CAF that can interact with E2F1-3 [145]. It has also been reported that HDACs can deacetylate and thus directly inactivate E2F transcription factors [145].

The binding of HPV16-E7 to HDAC is mediated by the Mi2β component of the nucleosomal remodelling and deacetylation complex (NURD), which can directly associate both E7 and HDAC. As shown by mutational studies, the zinc-finger domain of E7 is involved in this association, whereas mutations in pRb binding site do not abolish E7-HDAC interaction [23, 84]. It has also been evidenced that, while mutations in the LXCXE-binding site of pRb abolish the interaction with E7, an intact LXCXE cleft is not required for pRb interaction with HDACs, even though they also contain an LXCXE motif [146]. However, it has been shown that the binding of pRb to the HDACs is suppressed in presence of E7 [143]. Modulation of histone deacetylase activity by E7 could further contribute to de-repress E2F-responsive genes. Moreover, as already described in this chapter, E7 binding to HDAC seems to be involved in cdc25A up-regulation [47]. Most interestingly, E7 inability to associate with HDAC correlates with lack of growth-promoting activity. In fact, E7 mutants defective in HDAC binding are also deficient in transforming activity, suggesting a tight link between the oncogenic potential of this viral product and the binding to HDAC.

The modulation of HDACs exerted by E7 might have a further functional relevance in HPV pathogenesis as it appears to be responsible for the transcriptional regulation of immune-modulating molecules. In fact, recruitment of HDAC is involved in E7-induced silencing of interferon regulatory factor-1 (IRF-1), which has a central role in interferon signalling and immune surveillance [88]. In addition, a recent study has shown that HPV16-E7 recruits and/or stabilizes the HDACs on the major histocompatibility complex (MHC) class I promoter [147]. Deacetylation of the chromatin in correspondence of this promoter has been demonstrated to be specifically linked to E7 presence, as abrogation of the oncoprotein expression by siRNA also eliminates the HDACs from the class I promoter and results in histone acetylation. E7-induced chromatin remodelling through modulation of the HDACs could play a key role in down-regulating MHC class I molecules expression on the surface of HPV-positive cells. Therefore, E7-HDAC association might contribute to the mechanisms of escape from immune surveillance evolved by the HPV, reducing the recognition and elimination of HPV-infected cells by the immune system and favouring the establishment of a persistent infection.

Finally, the binding to HDACs has also been demonstrated to play a key role in mediating E7 function in the viral life cycle, since it seems to be necessary to maintain HPV episomal genomes and to extend KCs life span [23].

600kda Retinoblastoma Protein-Associated Factor (P600)

Identification of novel E7 targets has recently led to isolation of p600 as a cellular target of HPV16-E7 [148]. E7 N-terminus is required for the interaction with p600, though the binding appears to be pRb-independent. Data obtained in this study strongly suggest a role of p600 in E7 transforming activities. In fact, p600 depletion from HPV16-positive CaSki cells, as well from E6/E7-transformed NIH3T3 cells, partially suppresses the transformed phenotype. However, as LR-HPV E7 proteins also show the ability of associating p600, the relevance of this factor in the HPV-mediated transformation process must be assessed.

Transcription Factors (AP-1, TBP) and Transcriptional Coactivators (P300, P/CAF)

E7 has been reported to interfere with the activity of several transcription factors, such as AP-1 [82], TATA-box binding protein (TBP) [83], IRF-1 [88] and the forkhead transcription factor MPP2 [93].

The interaction of E7 with several components of the AP-1 transcription complex, including c-Jun, JunB, JunD and c-Fos, has been proven both *in vitro* and *in vivo* [82]. E7 ability of up-regulating AP-1 activity, which requires an intact zinc-finger domain, has been demonstrated and suggests relevance for E7 transforming properties. In fact, E7 inability of activating c-Jun-mediated transcription correlates to its inability to transform rat embryo fibroblasts [82]. In addition, a correlation has been noted between the ability of E7 mutants of transactivating c-Jun-dependent transcription and immortalizing primary keratinocytes, strongly suggesting a direct link between E7-driven transformation and modulation of AP-1 components. Therefore, the biological significance of this interaction is clear, although its exact contribution to E7 oncogenic potential is still at hypothesis level. It has been argued that the formation of E7-AP1 complexes may be important in the viral life cycle since AP-1 is also involved in regulating the transcription of E6/E7 from the viral genome and in keratinocyte differentiation [82].

E7 association with TBP has also been reported [83]. Interestingly, this interaction is influenced by the state of the E7-CKII target site, as phosphorylation at Ser 31 and Ser 32 of HPV16-E7 greatly enhances the binding to TBP. Data have been obtained regarding E7 ability of inhibiting TBP's DNA binding [149]. Association between the two proteins seems to be required for this inhibitory effect, which is increased upon E7 phosphorylation as well. The significance of TBP inhibition is still not clear, as a broad regulation of all TATA box-containing promoters is unlikely.

Further data have confirmed E7 capacity of transcriptional control, since its interaction with the transcriptional coactivators p300 and p300/CBP-associated factor (P/CAF) has also been evidenced [150, 151]. p300 displays a number of functions: it is an essential component of the transcription apparatus, possesses intrinsic HAT activity and can recruit other HATs, such as P/CAF. Numerous studies have demonstrated that this protein is essentially a tumor suppressor, involved in cell cycle control, differentiation and development. HPV16-E7 has been shown to interact with p300 and modulate its activity [150], which is, interestingly, a common feature of other viral oncoproteins, such as Ad E1A [152], SV40 T-Ag [27] and HPV-E6 [152]. Most importantly, interaction with p300 seems to be required for the transformation induced by these proteins [27, 152, 154]. However, if E7-p300 association plays a role in E7-mediated transformation is still debated, since also the product of LR-HPVs shows the ability of associating this HAT [150]. Both HPV16- and HPV11-E7 abolish p300 transcriptional function, even if to a different extent. In particular, E7 represses p300 ability to coactivate E2-dependent transcription [150]. Due to the role of E2 in regulating the expression of HPV oncoproteins, these data suggest a possible contribution of E7 to the regulation of its own transcription.

The association of E7 with P/CAF has also been described [151]. Importantly, E7 binding to this coactivator has been shown to be responsible for E7-induced repression of the interleukin-8 (IL-8) promoter [151], further confirming E7 ability of interfering with the expression of immuno-modulating molecules at transcriptional level. Subversion of the immune response may be important in evasion of HPV-infected cells from the immune surveillance, favoring a persistent infection and subsequent tumorigenesis (see below).

M2 Pyruvate Kinase (MP-K2)

Although most E7 targets are nuclear proteins, the cytolpasmic enzyme M2 pyruvate kinase (M2-PK) has also been recognized as a target of E7 [85]. This enzyme catalyses one of the limiting passages of glycolysis, i.e. the conversion of phosphoenolpyruvate (PEP) into pyruvate. E7 can influence the equilibrium between the two multimeric forms of the enzyme, which are characterized by different affinity for the substrate PEP [85]. As a consequence of this perturbed equilibrium, pyruvate production is decreased and the pool of phosphometabolytes upstream of PEP is expanded. The level of these metabolites, necessary for the biosynthesis of nucleotides, is usually low in quiescent cells. Therefore the interaction of E7 and M2-PK may have role in inducing the proliferation of the infected cells.

E7 has also been reported to affect the cellular carbohydrate metabolism through the direct interaction with the glycogen-degrading enzyme α-glucosidase [86]. This association promotes the activation of the enzyme catalytic activity *in vivo*, resulting in depletion of intracellular glycogen. Changes in carbohydrate metabolism are typical during malignant transformation. Significantly, they are frequently observed in early lesions of the cervix. Therefore, a potential link between E7 binding to α-glucosidase and E7-induced transformation may be hypothesized, though the precise role of the modulation of this enzyme is still not clear.

Subversion of p53 Function and the Apoptotic Pathway

Although in HPV-positive cells subversion of p53 functions is mainly ascribed to E6, interference with p53 activity by E7 has also been reported. Several studies have evidenced that E7 can overcome p53-induced growth arrest in response to cell damage [155, 156]. This phenomenon could depend on E7 capacity of inducing aberrant expression of cycA and cycE, as well as decreasing pRb levels and inactivating p53- responsive $p21^{CIP1/WAF1}$. Interestingly, E7-expressing cells show increased levels of p53 [157], suggesting that p53 degradation is perturbed. Indeed, in E7-positive cells p53 is not efficiently bound by the ubiquitine ligase MDM2, which normally regulates p53 steady-state level by targeting the tumor suppressor for proteasomal degradation [158]. As a result of p53 stabilization, the expression of E7 alone could induce apoptotic response [159]. However, as a consequence of E6/E7 synergistic and complementary effect, E7-induced apoptotic death of HPV-positive cells is bypassed by E6-mediated degradation of p53 and Bak [160].

E7 ability to decrease p53-dependent transcription in a transient transfection system has also been reported [161]. The binding to pRb seems not to be required for this activity while phosphorylation at the E7-CKII recognition site appears to be relevant. In fact, E7 phosphorylation by CKII increases E7 binding to p53, possibly mediated by TBP. Repression of p53 transcriptional activity is also induced by LR-HPV E7 proteins, suggesting the lack of a direct correlation between this property and HR-HPVs E7 transforming potential.

In spite of the described effects of E7 on p53 and the consequent apoptotic death E7 could induce in absence of E6, it is worth noting that E7 is a pleiomorphic protein as for apoptotic pathways. In fact, even if the majority of studies have evidenced E7 pro-apoptotic activity [162, 163], several data have indicated E7 as an anti-apoptotic protein. In a recent study, E6/E7 have been shown to inhibit TNF-mediated apoptosis in keratinocytes through up-regulation of the cellular inhibitor of apoptosis protein c-IAP2, though this activity seems to be mainly due to E6 [164]. Moreover, E7 has been reported to prevent TNF- and Fas-induced apoptosis in fibroblasts [165]. These contradictory results may be ascribed to different genetic backgrounds, cellular environments and experimental systems used, but also underline the complexity of E7 activities and effects.

E7 as an Inducer of Genomic Instability

Numerous recent studies have evidenced E7 ability to induce abnormal centrosomal duplication, which is a unique property of the HR-HPV protein [166]. The data obtained are consistent with the previous observations that E7 expression rapidly leads to aneuploidy and genomic instability [167]. In fact, as a result of defects in centriole assembly, multipolar mitoses occur and chromosomes do not symmetrically segregate during cell division. Indeed, tripolar mitotic figures are histopathological marks of HR-HPVs positive lesions, which typically contain irregular centrosome number [44, 166].

Abnormal centrosomal synthesis in HPV-positive cells seems to be independent of E7 ability to bind and degrade the "pocket proteins". In fact, pRb-, p107- and p130-deficient cells, which already show centrioles numeric irregularity, display a significant increase in this phenomenon upon HPV16-E7 expression [168]. Moreover, this activity is not associated with E7 ability to deregulate the expression of Cdc25, though its homologue in *Drosophila* is involved in centriole synthesis and maturation. The inactivation of $p21^{CIP1/WAF1}$ and $p27^{KIP1}$ has been suggested to play an important role in the subversion of centrosome homeostasis, since loss of $p21^{CIP1/WAF1}$ results in alteration of centrosome number [168]. Besides, the complexes cycE/cdk2 and cycA/cdk2 have been implicated in centrosome duplication in mammalian cells [170], thus it is conceivable that deregulation of this kinase contributes to the presence of supernumerary centrosomes. Indeed, cycE/cdk2 has been recently shown to be required for centriole overduplication from a single maternal template [171]. A recent study has demonstrated that HPV16-E7 possesses the ability to associate the gamma-tubulin component of centrosomes and that this interaction alters its recruitment to the centrosomes [172]. This study has also confirmed that E7-induced abnormal centrosomal synthesis relies on a "pocket proteins"-independent mechanism, since E7 association to gamma-tubulin does not involve the members of the retinoblastoma family [172].

As E7 proteins of LR-HPVs do not show the capacity of inducing irregular centrosome synthesis [166], it seems clear that this ability is greatly relevant in HR-HPV E7-driven transformation, as it confers an intrinsic element of genomic instability.

HPV-E7 also induces structural chromosome instability *in vitro* [166, 173]. Significantly, HPV-positive cancer cells frequently show structural chromosomal changes, such as translocations, deletions and amplifications [174]. Chromosome breaks could be caused by increased susceptibility to DNA damage [175] or defective DNA repair [173], presumably linked to loss of p53 and pRb function. Moreover, while cdc25A does not appear to be involved in E7-induced abnormal duplication of centrosomes, as already mentioned [168], its deregulation may be an important determinant of genome instability in E7-positive cells (see above). Importantly, disruption of chromosomal stability may ultimately promote integration of viral DNA in the host genome, an essential step in HPV-induced malignant transformation.

E7 in the Evasion of the Host Immune Response

Besides the numerous pathways of cell cycle control E7 is able to subvert, this oncoprotein appears to be essentially involved in HPV-induced immune suppression, which may partially account for the cases of persistent viral infection. In fact, even though most of the HPV infections in immunocompetent hosts are effectively controlled by the immune system, in a limited number of cases the virus establishes a persistent infection. A thorough understanding of the determinants that enable the HPV persistance is fundamental, since this persistance represents the main risk factor for the development of CC.

HPV has evolved specific mechanisms to escape from the host defences. Some of them are intrinsically linked to the virus life cycle. For instance, HPV reproduction in the upper layers of the epithelium, where the cells are rapidly and frequently eliminated by shedding, may determine a delayed/inefficient immune response. The superficial squamous epithelial cells are the only site where the viral gene products are expressed at high level; therefore, they are scarcely exposed to recognition by the cells of the immune system. Moreover, the ability of HPV proteins to modulate directly several components of the host defenses essentially contributes to determine that HPV-infected cells are not readily detected, and then eliminated, from the host immunity.

Several immune suppressor strategies have been developed by the HPV to prevent recognition of its antigens from the cells of the immune system. Among these, numerous have been ascribed to E7, which shows the ability of interfering with both innate and adaptive immune response.

A recent study has evidenced as HPV16 E6 and E7 are able to suppress the expression of the Toll-like receptor 9 (TLR9) [176]. This protein belongs to TLR family, which plays a key role at the interface between innate and adaptive immune response, recognizing and promoting the elimination of bacterial and viral pathogens. TLR9, in particular, is activated by dsDNA-derived CpG motifs. HPV16 and, at a lower extent, HPV18 oncoproteins have been shown to induce a significant decrease in TLR9 mRNA level, greatly compromising the functionality of TLR9 pathway [176]. Significantly, it has been reported that TLR9-specific staining is absent in cervical cancer biopsies in comparison with normal cervical tissue [176].

As already mentioned, E7 is involved in the deregulation of the expression and activity of immuno-modulating molecules, such as IL-8 [151]. In addition, E7 also subverts interferon (IFN) signalling pathways, by both affecting the expression and the function of type I interferons. These cytokines are essentially involved in host immune response since they stimulate the expression of MHC class I and II molecules, which in turn mediate the antigen recognition by cells of the immune system. Besides, IFNs also promote the expression of factors part of the machinery used for processing and presentation of the antigens on the cell surface. HPV16-E7 has been shown to inhibit multiple IFNα-inducible genes, due to its direct association with and sequestration of IRF-9 [91]. This interaction leads to the loss of IFN stimulatory gene factor-3 (ISGF3) transcription complex, which normally binds IFN responsive elements (ISRE) promoting the initiation of transcription of IFN-inducible genes.

As mentioned above, E7 has also been reported to impair IFNγ-induced IRF-1 transactivation activity, which is in turn involved in IFNβ gene expression [88, 89]. It is worth noting that IRF-1 also contributes to transcriptional activation of $p21^{CIP1/WAF1}$ [177], evidencing E7 potential ability of interfering with different processes, i.e. cell cycle control and immune response, inactivating a single multifunctional target.

HPV18-E7 has been shown to repress the transcription of the genes encoding the LMP2 subunit of the proteasomes involved in the production of the antigenic peptides, and TAP1, one of the components of the transporter associated with antigen presentation (TAP), that allows the peptides to be accumulated in the endoplasmic reticulum and loaded on MHC I molecules [178].

Several studies have also evidenced that, besides the effects achieved by inhibiting IFN pathways, E7 directly interferes with MHC expression, further decreasing the chance of HPV-positive cells to be recognized thus eliminated by T-cells. HPV-positive cervical carcinoma cells typically show low levels of MHC I molecules [179, 180]. As already mentioned, E7 down-regulates MHC class I genes in HPV16-positive cells through interaction with HDAC at level of MHC I heavy chain promoter [146]. E7 capacity of modulating MHC I expression has been confirmed by transfecting E7-siRNA in HPV-positive cells. In this system E7 knock out induces a significant increase in cell-surface MHC I molecules [181]. Moreover, overexpression of E7 in HPV-negative cells determines a marked reduction in MHC I molecules [181].

E7 has also been shown to influence the functionality of the adaptive immune system. For instance, E7 appears to repress the antigen-presenting function of dendritic cells (DCs) [182]. Indeed, DCs expressing E7 and pulsed with E7 are poor stimulators of E7-specific immunity [182].

Taken together, all these data demonstrate that E7 actively and directly inactivates the host immunity, partially but significantly contributing to HPV ability of escaping the immune surveillance. E7-mediated strategies of immune evasion may eventually lead to the virus persistence and ultimately contribute to the carcinogenic progression.

E7 as a Target for Cervical Cancer Therapy

HPV-induced cancers represent up to up to 15% of all cancer cases in women and up to 5% in men [183]. In particular, as previously mentioned, cervical cancer is the second most common cancer among women, accounting for one-fifth of all their cancer-related deaths [4].

Early detection of pre-cancerous lesions based on regular PAP-testing can markedly reduce cervical cancer death rates [184]. However, neither effective population screening programs nor optimal treatment are available in developing countries, where CC may represent the most prevalent cancer in women. Research on the biology of cervical cancer together with its viral etiology has recently resulted in the approval by the Food and Drug Administration (FDA) of an anti-HPV prophylactic vaccine based on the major capsid protein L1 of the High-Risk HPVs 16 and 18. This vaccine has been demonstrated to be effective in reducing the incidence of persistent HPV infections and HPV-associated lesions [185, 186]. Nevertheless, a significant impact of the vaccination on CC incidence will not be observed for at least a couple of decades. Moreover, despite the availability of an effective vaccine against HPV infection, a disparity in vaccination programs will probably occur due to the high cost of this new HPV vaccination for developing countries. Unfortunately, there is little evidence that the vaccine has therapeutic activity in CC patients and the survival benefits of the currently available anti-cancer therapies are not satisfying, especially in cases of advanced stage CC. The 5-year survival rate for CC has improved only by 4% over the last three decades; for patients with CC in advanced stage it is around 15-20%. The main therapeutic options, represented by surgery and radiation, are successful only if the tumor is detected at an early stage. In advanced stage, the efficacy of these therapies is less successful and the available chemotherapeutic agents are often only a palliative [187]. Radiotherapy and chemotherapy alone or in combination are used to eliminate potential residual cancer cells. However, these approaches can severely diminish the quality of life of the patients and are limited by the need of achieving therapeutically relevant drug concentrations. The toxicity associated with the high drug dose required to maintain a state of remission thus represents a major problem. Therefore, the pre-clinical and clinical studies focused on the development of strategies capable of providing an effective therapy for HPV-induced lesions while avoiding significant side effects are numerous. Since the results in terms of tumor regression are still not encouraging, a continuous effort to develop effective and less invasive therapies continues to be made.

The viral etiology of cervical cancer allows for choosing viral antigens as specific targets. HPV oncoproteins represent the most attractive candidate for a targeted cancer therapy, as they are the primary contributors to the development of malignant transformation. Due to E7 profound involvement in the oncogenesis, the inhibition of the function of this protein might be an effective approach to block the process of malignant transformation. E7 offers the unique advantage of being constitutively expressed only in HPV-positive cells and of being unrelated to the host proteins. Therefore E7 is a specific target, allowing selective targeting of only the pre-malignant and malignant cells, without any effect on normal cells.

As already mentioned in this chapter, E2 gene product can directly bind the HPV early promoter and repress the transcription of HPV oncogenes [16]. In addition, a recent study has reported that E2 can also regulate E7 activity [188]. In fact, E2 can suppress E7-induced transformation and this phenomenon appears to be independent from its ability to suppress E7 expression. Due to its function as a transcriptional repressor, introduction of E2 gene in HPV-positive cells has been evaluated as an approach to suppress the oncoproteins expression, thus to revert the transformed phenotype. An SV40-derived viral vector has been exploited to deliver the E2 ORF in HPV18-positive HeLa cells [18, 189]. Adenovirus-mediated delivery of BPV-E2 to CaSki cells has also been reported [190]. In both HPV16- and HPV18-positive cells, E2-mediated abrogation of the endogenous E6/E7 has been successfully achieved. Most interestingly, profound effects on DNA synthesis with rapid growth arrest and acquisition of the senescent phenotype have been observed [189, 190]. Upon ectopical E2 expression in HPV-positive cells, apoptosis induction has also been shown [191, 192], though this effect has been detected in HPV-negative cells as well [191]. Effects elicited by E2 have been partially reversed by constitutive expression of E6/E7, confirming that growth inhibition is due to the suppression of the oncogenes transcription mediated by E2 [9]. Importantly, the delivery of BPV-E2 into HeLa cells stably expressing either HPV16-E6 or E7 has allowed elucidation of the individual effects of the suppression of each oncoprotein [192]. This study has shown that E7 repression does not reactivate p53 pathway but does restore pRb pathway. Indeed, an increase in the levels of the hypophosphorylated form of pRb, p107 and p130 has been observed and suppression of the E2F-responsive genes for cycA and cdc25A has been detected. Moreover, E7 abrogation alone has been shown to be sufficient to inhibit telomerase activity. In this system, E7 repression efficiently triggered senescence, while appearance of apoptotic cells was limited. Conversely, E6 suppression has been demonstrated to reactivate p53 pathway and to induce both senescence and apoptosis [192].

Abundant data in the literature have proven that E7 inactivation/down-regulation at the post-transcriptional level is an effective approach for reversal of the malignant phenotype and/or killing of cervical cancer cells as well. Numerous strategies to suppress E7 activity by blocking the oncogene mRNA translation have been explored. For instance, HPV16-E7 expression knock out has been obtained by delivery of antisense (AS) RNA molecules, that can form a duplex with the RNA of the opposite strand leading to its degradation. ASRNA for HPV oncoproteins have been delivered to tumor cells by using adeno-associated virus [193, 194], herpes simplex virus-derived vectors [195] and replication-defective retrovirus [196]. All these delivery systems have been proven to be effective in specific suppression of the E6/E7 mRNA and the relative proteins. Up-regulation of pRb and down-regulation of E2F1-mediated transcription have been reported as a result of E7 knock-down by ASRNA [194, 196]. Most importantly, delivery of HPV16E7-ASRNA has been demonstrated to induce partial reversal of the malignant features of HPV16-positive CaSki cell line. Reduced migration and invasion capability of these cells *in vitro* have been reported [193]. Furthermore, antisense HPV16E7 constructs have been able to inhibit and/or retard the tumorigenicity of CaSki cells in nude mice [193, 194, 196]. ASRNA-mediated repression of both E6 and E7 products of HPV16 has also been reported to induce senescence and apoptosis of the human cancer SiHa cells [194, 197].

Successful silencing of HR-HPVE7 by small interfering RNA (siRNA) has also been documented [198-201]. In fact, siRNA-induced specific degradation of the target mRNA enables knock-down of the relative protein. Due to the bicistronic nature of the mRNA encodying E6 and E7, expression of E7 siRNA actually results in suppression of both the oncoproteins. However, some studies have reported the design of siRNA specifically targeting only E7 [198]. Growth inhibition and induction of senescence in HPV18-positive HeLa cells have been achieved by delivery of E6-E7siRNA [199, 200]. On the other hand, siRNA-mediated silencing of E7 in SiHa cells has been shown to induce apoptotic cell death [198]. Importantly, intra-tumor injection of specific siRNA has resulted in successful suppression of tumor growth *in vivo* [200]. As expected, E7 repression has also been described to rescue pRb [200].

Reduction of E7 protein level has also been achieved by using a lentiviral vector to deliver a short hairpin RNA (shRNA) against HPV18-E6 [202]. Induction of HPV18-positive cells senescence or specific cell death via apoptosis have been reported in this study. Additionally, inhibition of tumor growth *in vivo* has also been obtained upon systemic delivery of the shRNA in mice.

Among the strategies exploiting nucleic acids as potential therapeutic agents for HPV-associated cancer, the use of ribozymes (Rz) has also been documented, although it is mostly restricted to *in vitro* studies with very limited practical applications. Ribozymes possess a secondary structure that provides the molecule with enzymatic activity. Rz specifically designed for the transcripts of either HPV16- or HPV18-E6/E7 have been tested, showing the ability of effectively cleaving the target HPV-RNA *in vitro* [203-205]. These results have provided the basis to evaluate the validity of Rz-based therapeutic approaches. Indeed, some studies have reported the ribozymes to be effective in preventing E6/E7-mediated immortalization of human keratinocytes [205] and in inhibiting CaSki cells growth [206]. Interestingly, CaSki cells expressing specific ribozymes have revealed higher sensitivity to cisplatin and radiation, offering the possibility of using Rz to increase the efficacy of the current available therapies. However, a pre-clinical study failed to detect any anti-tumor effect in nude mice upon delivery of ribozyme genes by adeno-associated virus [207].

Since strategies based on RNA are limited by mRNA turnover and can induce non-specific effects (siRNA), other approaches aimed at directly impairing the activity of E7 at protein level have been explored. For instance, the use of peptide aptamers to block E7 transforming potential has been assessed [208]. This study has demonstrated that 20-mer peptides selected against the CR3 of HPV16-E7 can modulate the oncoprotein activity in living cells. In particular, some of the isolated peptides have shown the ability of inhibiting the growth of NIH3T3E7 and CaSki cells, through an increase in the apoptotic rate.

Suppression of E7-mediated proliferation has also been achieved by intracellular expression of specific antibodies in single-chain format (scFv) [209]. Anti-E7 antibodies selected by phage display technology have been shown to reduce DNA synthesis when expressed in the nucleus and endoplasmic reticulum of HPV16-positive SiHa cells. In comparison with other approaches, therapeutic strategies based on intracellular antibodies ("intrabodies") offer the advantage of potentially blocking intermolecular interactions, which are fundamental for oncoproteins transforming activity. Indeed, in a recent paper specific "intrabodies" have demonstrated the ability to suppress E6-mediated degradation of p53, which occurs upon the formation of a trimeric complex involving E6, p53 and the cellular protein E6AP [210].

Finally, E6 and E7 might be exploited as targets for anti-cancer drug discovery since their domains as well as many of the intermolecular interactions that contribute to their oncogenic potential are identified. In fact, the development of new chemotherapeutics to manage CC should be directed by the knowledge of the mechanisms underlying the development of this tumor, even if partial. Therefore, the effort to develop new anti-HPV chemotherapeutics designed to affect the structure and/or the function of the oncoproteins should be pursued.

Several compounds, such as sulindac, cidofovir, curcumin and ursolic acid, among others, have been reported to be effective in suppressing the level of E6/E7 in HPV-positive cells [211-215]. These drugs have also shown a certain ability to inhibit the growth of cancer cells and to promote apoptosis. However, none of these compounds has been specifically designed for binding to the HPV oncoproteins and most of the time their effects are a mere result of serendipity.

Discovery of new anti-HPV compounds could greatly benefit from the knowledge of the specific functional relationship between the target domains and activity. Indeed, studies aimed to identify and characterize compounds that interfere with E6 binding to zinc have been reported [216, 217]. Importantly, several zinc-ejecting compounds have shown the ability to inhibit E6 binding to some of its target, such as E6AP and E6-binding protein (E6BP). Most significantly, they have been able to promote specific death of HPV-positive cells [217]. These studies have demonstrated that targeting a specific domain associated with a known and relevant biological function can be effective in suppressing the target activity. Moreover, they have shown that small molecules represent a valuable tool in inhibiting intermolecular interactions, which are often the basis of tumorigenic processes. However, the approach of screening compounds for their potential ability of interacting with E7 and interfering with its activity has not been reported yet. Computational biology may have an important role in this field, as the use of computational methods has been proven to be effective for isolating specific inhibitors of the molecules of interest [218-221] and potential anti-cancer drugs [222, 223].

Conclusion

The oncogenic potential of E7 protein of High-Risk Human Papillomaviruses has been extensively documented over the years. Its ability to deregulate numerous cellular modulators has evidenced how this oncoprotein can affect so many fundamental processes, such as cell cycle control, differentiation, apoptosis and immune response. Since the malignant progression represents an unintended consequence of the viral infection, it is obvious as the deregulation of these processes has the main significance of allowing the virus replication, which is, ultimately, the central function of E7. However, it is also clear as the extensive deregulation of E7 expression and the continuous subversion of crucial cellular pathways, for instance following the integration of the viral DNA in the host genome, essentially contribute to HPV-mediated carcinogenesis.

Although the list of E7 cellular targets keeps growing and new insights in the mechanisms of E7-induced transformation are emerging, the critical contribution of pRb inactivation and destabilization to E7 transforming properties is unquestionable. The loss of pRb negative modulation on the cell cycle certainly has a role in promoting deregulated proliferation of the infected cells through release of E2F-mediated transcription. Nevertheless, modulation of other cellular proteins has a biological relevance in the carcinogenic process, as evidenced by the capability of pRb binding-defective mutants to induce immortalization and of non-transforming mutants of E7 to still bind pRb. Numerous cellular factors targeted by E7, including the other "pocket proteins" p107 and p130, as well as cyclins and cyclin-dependent kinases, inhibitors of cdks, histone deacetylases, transcription factors and glycolytic enzymes, might be important determinants for E7 oncogenic potential. For instance, the property of E7 protein of the High-Risk HPVs to induce structural and numerical alterations of the cellular genome seems to be dependent on the deregulation of proteins other than pRb, though their exact contribution is still unclear. A thorough understanding of the effects induced by the additional associations established by E7, as well as of the underlying mechanisms, will clarify how extensively they contribute to the transformation. Regardless, it is evident that E7-induced carcinogenesis results from complex and mutually related events. It appears that E7-induced subversion of the host defenses, which essentially contributes to HPV ability to escape from the immune surveillance, ultimately leads to a condition that promotes the carcinogenic progression. In fact, together with other factors, it determines a persistent infection, which allows a prolonged expression of E7 and thus a situation of sustained oncogenic stimuli. Notably, E7-induced genetic changes clearly play a key role in the carcinogenesis, since they confer an endogenous element of instability that likely promotes the malignant conversion of the infected cells.

The concomitant expression of E6 and E7 in HPV-infected cells obviously determines a more elaborate scenario, in which the interplay between the two oncoproteins through a refined and specific "choice" of the respective targets ultimately establishes the conditions for the malignant progression.

It is worth noting that E7 ability to interact with multiple partners is really remarkable, considering that this protein contains just about 100 amino acids. However, as new data on E7 are emerging, this property seems to be less surprising. In fact, E7 has been described as an "intrinsically disordered" or "natively unfolded" protein. This "unstructured" aspect may determine a flexible and dynamic conformation that in turns may account for the extreme functional plasticity of this oncoprotein. Interestingly, in a recent study E7 proteins of High-Risk HPVs have been reported to be more disordered than E7 proteins of Low-Risk HPVs both in the N- and C-terminus, suggesting that this element of disorder might have a role with respect to the different biological function and oncogenic potential [61].

Finally, the essential role of E7 in HPV-mediated transformation and its features as a tumor-specific antigen make this protein an ideal target for the therapy of HPV-associated lesions. All the approaches to knock out E7 both at transcriptional and post-transcriptional level, mentioned in this chapter, have major therapeutic potential for the treatment of cervical cancer. However, it must be emphasized that most of the successful results obtained by these strategies are restricted to *in vitro* applications and pre-clinical models and that there are still several challenges to their clinical potential. For instance, several important issues remain unsolved in gene therapy, such as the development of a safe and effective delivery system. The validation of new potential chemotherapeutics also presents important barriers. A great effort is required to identify cancer-specific drugs, to obtain their proper delivery to the targets cells and to ensure the administration of the therapeutic agent in relevant and effective dose without toxicity. Nonetheless, the scenario of therapeutic options for cervical cancer will benefit from a complete understanding of the mechanisms of E7-mediated carcinogenesis, together with the development of new technologies for the isolation, pre-clinical development and safe delivery of therapeutic molecules.

Reviewed by Prof. Alfred B. Jenson, MD (James Graham Brown Cancer Center, University of Louisville), Dr. Douglas C. Dean, PhD (Department of Ophthalmology, University of Louisville) Dr. Kenneth E. Palmer, PhD (Department of Pharmacology and Toxicology, University of Louisville).

Acknowledgments

The author gratefully acknowledges Miss Amanda B. Lasnik for reviewing the English of the manuscript.

References

[1] de Villiers, EM; Gunst, K; Stein, H; Scherubl, H. Esophageal squamous cell cancer in patients with head and neck cancer: Prevalence of human papillomavirus DNA sequences. *International journal of cancer*, 2004, Mar 20, 109(2), 253-8.

[2] Bosch, FX; de Sanjose, S. Human papillomavirus in cervical cancer. *Current oncology reports*, 2002, Mar, 4(2), 175-83.

[3] Walboomers, JM; Jacobs, MV; Manos, MM; Bosch, FX; Kummer, JA; Shah, KV; et al. Human papillomavirus is a necessary cause of invasive cervical cancer worldwide. The *Journal of pathology*, 1999, Sep, 189(1), 12-9.

[4] Parkin, DM; Bray, F; Ferlay, J; Pisani, P. Global cancer statistics, 2002. CA: a *cancer journal for clinicians*, 2005, Mar-Apr, 55(2), 74-108.

[5] Hawley-Nelson, P; Vousden, KH; Hubbert, NL; Lowy, DR; Schiller, JT. HPV16 E6 and E7 proteins cooperate to immortalize human foreskin keratinocytes. *The EMBO journal*, 1989, Dec 1, 8(12), 3905-10.

[6] Munger, K; Phelps, WC; Bubb, V; Howley, PM; Schlegel, R. The E6 and E7 genes of the human papillomavirus type 16 together are necessary and sufficient for transformation of primary human keratinocytes. *Journal of virology.* 1989 Oct, 63(10), 4417-21.

[7] Durst, M; Gallahan, D; Jay, G; Rhim, JS. Glucocorticoid-enhanced neoplastic transformation of human keratinocytes by human papillomavirus type 16 and an activated ras oncogene. *Virology*, 1989, Dec, 173(2), 767-71.

[8] Goodwin, EC; DiMaio, D. Repression of human papillomavirus oncogenes in HeLa cervical carcinoma cells causes the orderly reactivation of dormant tumor suppressor pathways. *Proceedings of the National Academy of Sciences of the United States of America*, 2000, Nov 7, 97(23), 12513-8.

[9] Wells, SI; Francis, DA; Karpova, AY; Dowhanick, JJ; Benson, JD; Howley, PM. Papillomavirus E2 induces senescence in HPV-positive cells via pRB- and p21(CIP)-dependent pathways. *The EMBO journal*, 2000, Nov 1, 19(21), 5762-71.

[10] Smotkin, D; Wettstein, FO. Transcription of human papillomavirus type 16 early genes in a cervical cancer and a cancer-derived cell line and identification of the E7 protein. *Proceedings of the National Academy of Sciences of the United States of America*, 1986, Jul, 83(13), 4680-4.

[11] Androphy, EJ; Lowy, DR; Schiller, JT. Bovine papillomavirus E2 trans-activating gene product binds to specific sites in papillomavirus DNA. *Nature.* 1987, Jan 1-7, 325(6099), 70-3.

[12] Baker, CC; Phelps, WC; Lindgren, V; Braun, MJ; Gonda, MA; Howley, PM. Structural and transcriptional analysis of human papillomavirus type 16 sequences in cervical carcinoma cell lines. *Journal of virology*, 1987, Apr, 61(4), 962-71.

[13] Cullen, AP; Reid, R; Campion, M; Lorincz, AT. Analysis of the physical state of different human papillomavirus DNAs in intraepithelial and invasive cervical neoplasm. *Journal of virology*, 1991, Feb, 65(2), 606-12.

[14] Fujii, T; Masumoto, N; Saito, M; Hirao, N; Niimi, S; Mukai, M; et al. Comparison between in situ hybridization and real-time PCR technique as a means of detecting the integrated form of human papillomavirus 16 in cervical neoplasia. *Diagn.Mol. Pathol*, 2005, Jun, 14(2), 103-8.

[15] Hudelist, G; Manavi, M; Pischinger, KI; Watkins-Riedel, T; Singer, CF; Kubista, E; et al. Physical state and expression of HPV DNA in benign and dysplastic cervical tissue: different levels of viral integration are correlated with lesion grade. *Gynecologic oncology*, 2004, Mar, 92(3), 873-80.

[16] McBride, AA; Romanczuk, H; Howley, PM. The papillomavirus E2 regulatory proteins. *The Journal of biological chemistry*, 1991, Oct 5, 266(28), 18411-4.

[17] Dowhanick, JJ; McBride, AA; Howley, PM. Suppression of cellular proliferation by the papillomavirus E2 protein. *Journal of virology*, 1995, Dec, 69(12), 7791-9.

[18] Hwang, ES; Riese, DJ; 2nd; Settleman, J; Nilson, LA; Honig, J; Flynn, S; et al. Inhibition of cervical carcinoma cell line proliferation by the introduction of a bovine papillomavirus regulatory gene. *Journal of virology*, 1993, Jul, 67(7), 3720-9.

[19] Corden, SA; Sant-Cassia, LJ; Easton, AJ; Morris, AG. The integration of HPV-18 DNA in cervical carcinoma. *Mol. Pathol*, 1999, Oct, 52(5), 275-82.

[20] Choo, KB; Pan, CC; Han, SH. Integration of human papillomavirus type 16 into cellular DNA of cervical carcinoma: preferential deletion of the E2 gene and invariable retention of the long control region and the E6/E7 open reading frames. *Virology*, 1987, Nov, 161(1), 259-61.

[21] Jeon, S; Lambert, PF. Integration of human papillomavirus type 16 DNA into the human genome leads to increased stability of E6 and E7 mRNAs: implications for cervical carcinogenesis. *Proceedings of the National Academy of Sciences of the United States of America*, 1995, Feb 28, 92(5), 1654-8.

[22] Flores, ER; Allen-Hoffmann, BL; Lee, D; Lambert, PF. The human papillomavirus type 16 E7 oncogene is required for the productive stage of the viral life cycle. *Journal of virology*, 2000, Jul, 74(14), 6622-31.

[23] Longworth, MS; Laimins, LA. The binding of histone deacetylases and the integrity of zinc finger-like motifs of the E7 protein are essential for the life cycle of human papillomavirus type 31. *Journal of virology*, 2004, Apr, 78(7), 3533-41.

[24] Collins, AS; Nakahara, T; Do, A; Lambert, PF. Interactions with pocket proteins contribute to the role of human papillomavirus type 16 E7 in the papillomavirus life cycle. *Journal of virology*, 2005, Dec, 79(23), 14769-80.

[25] Dyson, N; Guida, P; McCall, C; Harlow, E. Adenovirus E1A makes two distinct contacts with the retinoblastoma protein. *Journal of virology*, 1992, Jul, 66(7), 4606-11.

[26] Ben-Israel, H; Kleinberger, T. Adenovirus and cell cycle control. *Front Biosci*, 2002, May 1, 7, d1369-95.

[27] Ali, SH; DeCaprio, JA. Cellular transformation by SV40 large T antigen, interaction with host proteins. *Seminars in cancer biology*, 2001, Feb, 11(1), 15-23.

[28] Dyson, N; Howley, PM; Munger, K; Harlow, E. The human papilloma virus-16 E7 oncoprotein is able to bind to the retinoblastoma gene product. *Science* (New York, NY. 1989, Feb 17, 243(4893), 934-7.

[29] Werness, BA; Levine, AJ; Howley, PM. Association of human papillomavirus types 16 and 18 E6 proteins with p53. Science (New York, NY. 1990, Apr 6, 248(4951), 76-9.

[30] Scheffner, M; Werness, BA; Huibregtse, JM; Levine, AJ; Howley, PM. The E6 oncoprotein encoded by human papillomavirus types 16 and 18 promotes the degradation of p53. *Cell*, 1990, Dec 21, 63(6), 1129-36.

[31] Huibregtse, JM; Scheffner, M; Howley, PM. A cellular protein mediates association of p53 with the E6 oncoprotein of human papillomavirus types 16 or 18. *The EMBO journal*, 1991, Dec, 10(13), 4129-35.

[32] Accardi, R; Dong, W; Smet, A; Cui, R; Hautefeuille, A; Gabet, AS; et al. Skin human papillomavirus type 38 alters p53 functions by accumulation of deltaNp73. *EMBO reports*, 2006, Mar, 7(3), 334-40.

[33] Halbert, CL; Demers, GW; Galloway, DA. The E6 and E7 genes of human papillomavirus type 6 have weak immortalizing activity in human epithelial cells. *Journal of virology*, 1992, Apr, 66(4), 2125-34.

[34] Vousden, KH; Doniger, J; DiPaolo, JA; Lowy, DR. The E7 open reading frame of human papillomavirus type 16 encodes a transforming gene. *Oncogene research*, 1988, Sep, 3(2), 167-75.

[35] Matlashewski, G; Schneider, J; Banks, L; Jones, N; Murray, A; Crawford, L. Human papillomavirus type 16 DNA cooperates with activated ras in transforming primary cells. *The EMBO journal*, 1987, Jun, 6(6), 1741-6.

[36] Phelps, WC; Yee, CL; Munger, K; Howley, PM. The human papillomavirus type 16 E7 gene encodes transactivation and transformation functions similar to those of adenovirus E1A. *Cell*, 1988, May 20, 53(4), 539-47.

[37] Liu, X; Disbrow, GL; Yuan, H; Tomaic, V; Schlegel, R. Myc and HPV-16 E7 genes cooperate to immortalize human keratinocytes. *Journal of virology*, 2007, Sep 5.

[38] Oh, ST; Kyo, S; Laimins, LA. Telomerase activation by human papillomavirus type 16 E6 protein: induction of human telomerase reverse transcriptase expression through Myc and GC-rich Sp1 binding sites. *Journal of virology*, 2001, Jun, 75(12), 5559-66.

[39] Veldman, T; Liu, X; Yuan, H; Schlegel, R. Human papillomavirus E6 and Myc proteins associate in vivo and bind to and cooperatively activate the telomerase reverse transcriptase promoter. *Proceedings of the National Academy of Sciences of the United States of America*, 2003, Jul 8, 100(14), 8211-6.

[40] Halbert, CL; Demers, GW; Galloway, DA. The E7 gene of human papillomavirus type 16 is sufficient for immortalization of human epithelial cells. *Journal of virology*, 1991, Jan, 65(1), 473-8.

[41] Wazer, DE; Liu, XL; Chu, Q; Gao, Q; Band, V. Immortalization of distinct human mammary epithelial cell types by human papilloma virus 16 E6 or E7. *Proceedings of the National Academy of Sciences of the United States of America*, 1995, Apr 25, 92(9), 3687-91.

[42] Herber, R; Liem, A; Pitot, H; Lambert, PF. Squamous epithelial hyperplasia and carcinoma in mice transgenic for the human papillomavirus type 16 E7 oncogene. *Journal of virology*, 1996, Mar, 70(3), 1873-81.

[43] Balsitis, SJ; Sage, J; Duensing, S; Munger, K; Jacks, T; Lambert, PF. Recapitulation of the effects of the human papillomavirus type 16 E7 oncogene on mouse epithelium by somatic Rb deletion and detection of pRb-independent effects of E7 in vivo. *Molecular and cellular biology*, 2003, Dec, 23(24), 9094-103.

[44] Riley, RR; Duensing, S; Brake, T; Munger, K; Lambert, PF; Arbeit, JM. Dissection of human papillomavirus E6 and E7 function in transgenic mouse models of cervical carcinogenesis. *Cancer research*, 2003, Aug 15, 63(16), 4862-71.

[45] Demers, GW; Espling, E; Harry, JB; Etscheid, BG; Galloway, DA. Abrogation of growth arrest signals by human papillomavirus type 16 E7 is mediated by sequences required for transformation. *Journal of virology*, 1996, Oct, 70(10), 6862-9.

[46] Zerfass, K; Schulze, A; Spitkovsky, D; Friedman, V; Henglein, B; Jansen-Durr, P. Sequential activation of cyclin E and cyclin A gene expression by human papillomavirus type 16 E7 through sequences necessary for transformation. *Journal of virology*, 1995, Oct, 69(10), 6389-99.

[47] Nguyen, DX; Westbrook, TF; McCance, DJ. Human papillomavirus type 16 E7 maintains elevated levels of the cdc25A tyrosine phosphatase during deregulation of cell cycle arrest. *Journal of virology*, 2002, Jan, 76(2), 619-32.

[48] Villa, LL; Vieira, KB; Pei, XF; Schlegel, R. Differential effect of tumor necrosis factor on proliferation of primary human keratinocytes and cell lines containing human papillomavirus types 16 and 18. *Molecular carcinogenesis*, 1992, 6(1), 5-9.

[49] Basile, JR; Zacny, V; Munger, K. The cytokines tumor necrosis factor-alpha (TNF-alpha) and TNF-related apoptosis-inducing ligand differentially modulate proliferation and apoptotic pathways in human keratinocytes expressing the human papillomavirus-16 E7 oncoprotein. *The Journal of biological chemistry*, 2001, Jun 22, 276(25), 22522-8.

[50] Pietenpol, JA; Stein, RW; Moran, E; Yaciuk, P; Schlegel, R; Lyons, RM; et al. TGF-beta 1 inhibition of c-myc transcription and growth in keratinocytes is abrogated by viral transforming proteins with pRB binding domains. *Cell*, 1990, Jun 1, 61(5), 777-85.

[51] Barbosa, MS; Edmonds, C; Fisher, C; Schiller, JT; Lowy DR; Vousden, KH. The region of the HPV E7 oncoprotein homologous to adenovirus E1a and Sv40 large T antigen contains separate domains for Rb binding and casein kinase II phosphorylation. *The EMBO journal*, 1990, Jan, 9(1), 153-60.

[52] Massimi, P; Banks, L. Differential phosphorylation of the HPV-16 E7 oncoprotein during the cell cycle. *Virology*, 2000, Oct 25, 276(2), 388-94.

[53] McIntyre, MC; Frattini, MG; Grossman, SR; Laimins, LA. Human papillomavirus type 18 E7 protein requires intact Cys-X-X-Cys motifs for zinc binding, dimerization, and transformation but not for Rb binding. *Journal of virology*, 1993, Jun, 67(6), 3142-50.

[54] Ohlenschlager, O; Seiboth, T; Zengerling, H; Briese, L; Marchanka, A; Ramachandran, R; et al. Solution structure of the partially folded high-risk human papilloma virus 45 oncoprotein E7. *Oncogene*, 2006, Sep 28, 25(44), 5953-9.

[55] Watanabe, S; Sato, H; Furuno, A; Yoshiike, K. Changing the spacing between metal-binding motifs decreases stability and transforming activity of the human papillomavirus type 18 E7 oncoprotein. *Virology*, 1992, Oct, 190(2), 872-5.

[56] Chellappan, S; Kraus, VB; Kroger, B; Munger, K; Howley, PM; Phelps, WC; et al. Adenovirus E1A, simian virus 40 tumor antigen, and human papillomavirus E7 protein share the capacity to disrupt the interaction between transcription factor E2F and the retinoblastoma gene product. *Proceedings of the National Academy of Sciences of the United States of America*, 1992, May 15, 89(10), 4549-53.

[57] Clements, A; Johnston, K; Mazzarelli, JM; Ricciardi, RP; Marmorstein, R. Oligomerization properties of the viral oncoproteins adenovirus E1A and human papillomavirus E7 and their complexes with the retinoblastoma protein. *Biochemistry*, 2000, Dec 26, 39(51), 16033-45.

[58] Jewers, RJ; Hildebrandt, P; Ludlow, JW; Kell, B; McCance, DJ. Regions of human papillomavirus type 16 E7 oncoprotein required for immortalization of human keratinocytes. *Journal of virology*, 1992, Mar, 66(3), 1329-35.

[59] Pahel, G; Aulabaugh, A; Short, SA; Barnes, JA; Painter, GR; Ray, P; et al. Structural and functional characterization of the HPV16 E7 protein expressed in bacteria. *The Journal of biological chemistry*, 1993, Dec 5, 268(34), 26018-25.

[60] Alonso, LG; Garcia-Alai, MM; Nadra, AD; Lapena, AN; Almeida, FL; Gualfetti, P; et al. High-risk (HPV16) human papillomavirus E7 oncoprotein is highly stable and extended, with conformational transitions that could explain its multiple cellular binding partners. *Biochemistry*, 2002, Aug 20, 41(33), 10510-8.

[61] Uversky, VN; Roman, A; Oldfield, CJ; Dunker, AK. Protein intrinsic disorder and human papillomaviruses: increased amount of disorder in E6 and E7 oncoproteins from high risk HPVs. *Journal of proteome research*, 2006, Aug, 5(8), 1829-42.

[62] Smotkin, D; Wettstein, FO. The major human papillomavirus protein in cervical cancers is a cytoplasmic phosphoprotein. *J.Virol*, 1987, May, 61(5), 1686-9.

[63] Selvey, LA; Dunn, LA; Tindle, RW; Park, DS; Frazer, IH. Human papillomavirus (HPV) type 18 E7 protein is a short-lived steroid-inducible phosphoprotein in HPV-transformed cell lines. *The Journal of general virology*, 1994, Jul, 75 (Pt 7), 1647-53.

[64] Reinstein, E; Scheffner, M; Oren, M; Ciechanover, A; Schwartz, A. Degradation of the E7 human papillomavirus oncoprotein by the ubiquitin-proteasome system: targeting via ubiquitination of the N-terminal residue. *Oncogene*, 2000, Nov 30, 19(51), 5944-50.

[65] Oh, KJ; Kalinina, A; Wang, J; Nakayama, K; Nakayama, KI; Bagchi, S. The papillomavirus E7 oncoprotein is ubiquitinated by UbcH7 and Cullin 1- and Skp2-containing E3 ligase. *Journal of virology*, 2004, May, 78(10), 5338-46.

[66] Zatsepina, O; Braspenning, J; Robberson, D; Hajibagheri, MA; Blight, KJ; Ely, S; et al. The human papillomavirus type 16 E7 protein is associated with the nucleolus in mammalian and yeast cells. *Oncogene*, 1997, Mar 13, 14(10), 1137-45.

[67] Sato, H; Watanabe, S; Furuno, A; Yoshiike, K. Human papillomavirus type 16 E7 protein expressed in Escherichia coli and monkey COS-1 cells, immunofluorescence detection of the nuclear E7 protein. *Virology*, 1989, May, 170(1), 311-5.

[68] Smith-McCune, K; Kalman, D; Robbins, C; Shivakumar, S; Yuschenkoff, L; Bishop, JM. Intranuclear localization of human papillomavirus 16 E7 during transformation and preferential binding of E7 to the Rb family member p130. *Proceedings of the National Academy of Sciences of the United States of America*, 1999, Jun 8, 96(12), 6999-7004.

[69] Dyson, N; Guida, P; Munger, K; Harlow, E. Homologous sequences in adenovirus E1A and human papillomavirus E7 proteins mediate interaction with the same set of cellular proteins. *Journal of virology*, 1992, Dec, 66(12), 6893-902.

[70] Cobrinik, D. Pocket proteins and cell cycle control. *Oncogene*, 2005, Apr 18, 24(17), 2796-809.

[71] Rowland, BD; Bernards, R. Re-evaluating cell-cycle regulation by E2Fs. *Cell*, 2006, Dec 1, 127(5), 871-4.

[72] Cam, H; Dynlacht, BD. Emerging roles for E2F: beyond the G1/S transition and DNA replication. *Cancer cell*, 2003, Apr, 3(4), 311-6.

[73] Harbour, JW; Dean, DC. Chromatin remodeling and Rb activity. *Current opinion in cell biology*, 2000, Dec, 12(6), 685-9.

[74] Godefroy, N; Lemaire, C; Mignotte, B; Vayssiere, JL. p53 and Retinoblastoma protein (pRb): a complex network of interactions. *Apoptosis*, 2006, May, 11(5), 659-61.

[75] Boyer, SN; Wazer, DE; Band, V. E7 protein of human papilloma virus-16 induces degradation of retinoblastoma protein through the ubiquitin-proteasome pathway. *Cancer* research, 1996, Oct 15, 56(20), 4620-4.

[76] Huh, K; Zhou, X; Hayakawa, H; Cho, JY; Libermann, TA; Jin, J; et al. Human papillomavirus type 16 e7 oncoprotein associates with the cullin 2 ubiquitin ligase complex, which contributes to degradation of the retinoblastoma tumor suppressor. *Journal of virology*, 2007, Sep, 81(18), 9737-47.

[77] Gonzalez, SL; Stremlau, M; He, X; Basile, JR; Munger, K. Degradation of the retinoblastoma tumor suppressor by the human papillomavirus type 16 E7 oncoprotein is important for functional inactivation and is separable from proteasomal degradation of E7. *Journal of virology*, 2001, Aug, 75(16), 7583-91.

[78] Zhang, B; Chen, W; Roman, A. The E7 proteins of low- and high-risk human papillomaviruses share the ability to target the pRB family member p130 for degradation. *Proceedings of the National Academy of Sciences of the United States of America*, 2006, Jan 10, 103(2), 437-42.

[79] Funk, JO; Waga, S; Harry, JB; Espling, E; Stillman, B; Galloway, DA. Inhibition of CDK activity and PCNA-dependent DNA replication by p21 is blocked by interaction with the HPV-16 E7 oncoprotein. *Genes and development*, 1997, Aug 15, 11(16), 2090-100.

[80] Jones, DL; Alani, RM; Munger, K. The human papillomavirus E7 oncoprotein can uncouple cellular differentiation and proliferation in human keratinocytes by abrogating p21Cip1-mediated inhibition of cdk2. *Genes and development.* 1997, Aug, 15, 11(16), 2101-11.

[81] Zerfass-Thome, K; Zwerschke, W; Mannhardt, B; Tindle, R; Botz, JW; Jansen-Durr, P. Inactivation of the cdk inhibitor p27KIP1 by the human papillomavirus type 16 E7 oncoprotein. *Oncogene*, 1996, Dec 5, 13(11), 2323-30.

[82] Antinore, MJ; Birrer, MJ; Patel, D; Nader, L; McCance, DJ. The human papillomavirus type 16 E7 gene product interacts with and trans-activates the AP1 family of transcription factors. *The EMBO journal*, 1996, Apr 15, 15(8), 1950-60.

[83] Massimi, P; Pim, D; Storey, A; Banks, L. HPV-16 E7 and adenovirus E1a complex formation with TATA box binding protein is enhanced by casein kinase II phosphorylation. *Oncogene*, 1996, Jun 6, 12(11), 2325-30.

[84] Brehm, A; Nielsen, SJ; Miska, EA; McCance, DJ; Reid, JL; Bannister, AJ; et al. The E7 oncoprotein associates with Mi2 and histone deacetylase activity to promote cell growth. *The EMBO journal*, 1999, May 4, 18(9), 2449-58.

[85] Zwerschke, W; Mazurek, S; Massimi, P; Banks, L; Eigenbrodt, E; Jansen-Durr, P. Modulation of type M2 pyruvate kinase activity by the human papillomavirus type 16 E7 oncoprotein. *Proceedings of the National Academy of Sciences of the United States of America*, 1999, Feb 16, 96(4), 1291-6.

[86] Zwerschke, W; Mannhardt, B; Massimi, P; Nauenburg, S; Pim, D; Nickel, W; et al. Allosteric activation of acid alpha-glucosidase by the human papillomavirus E7 protein. *The Journal of biological chemistry*, 2000, Mar 31, 275(13), 9534-41.

[87] Berezutskaya, E; Bagchi, S. The human papillomavirus E7 oncoprotein functionally interacts with the S4 subunit of the 26 S proteasome. *The Journal of biological chemistry*, 1997, Nov 28, 272(48), 30135-40.

[88] Park, JS; Kim, EJ; Kwon, HJ; Hwang, ES; Namkoong, SE; Um, SJ. Inactivation of interferon regulatory factor-1 tumor suppressor protein by HPV E7 oncoprotein. Implication for the E7-mediated immune evasion mechanism in cervical carcinogenesis. *The Journal of biological chemistry*, 2000, Mar 10, 275(10), 6764-9.

[89] Perea, SE; Massimi, P; Banks, L. Human papillomavirus type 16 E7 impairs the activation of the interferon regulatory factor-1. *International journal of molecular medicine*, 2000, Jun, 5(6), 661-6.

[90] Mazzarelli, JM; Atkins, GB; Geisberg, JV; Ricciardi, RP. The viral oncoproteins Ad5 E1A, HPV16 E7 and SV40 TAg bind a common region of the TBP-associated factor-110. *Oncogene*, 1995, Nov 2, 11(9), 1859-64.

[91] Barnard, P; McMillan, NA. The human papillomavirus E7 oncoprotein abrogates signaling mediated by interferon-alpha. *Virology*, 1999, Jul 5, 259(2), 305-13.

[92] Rey, O; Lee, S; Baluda, MA; Swee, J; Ackerson, B; Chiu, R; et al. The E7 oncoprotein of human papillomavirus type 16 interacts with F-actin in vitro and in vivo. *Virology*, 2000, Mar 15, 268(2), 372-81.

[93] Luscher-Firzlaff, JM; Westendorf, JM; Zwicker, J; Burkhardt, H; Henriksson, M; Muller, R; et al. Interaction of the fork head domain transcription factor MPP2 with the human papilloma virus 16 E7 protein, enhancement of transformation and transactivation. *Oncogene*, 1999, Oct 7, 18(41), 5620-30.

[94] Schilling, B; De-Medina, T; Syken, J; Vidal, M; Munger, K. A novel human DnaJ protein, hTid-1, a homolog of the Drosophila tumor suppressor protein Tid56, can interact with the human papillomavirus type 16 E7 oncoprotein. *Virology*, 1998, Jul 20, 247(1), 74-85.

[95] Scheffner, M. Ubiquitin, E6-AP, and their role in p53 inactivation. *Pharmacology and therapeutics*, 1998, Jun, 78(3), 129-39.

[96] Mittnacht, S. Control of pRB phosphorylation. *Current opinion in genetics and development*, 1998, Feb, 8(1), 21-7.

[97] Munger, K; Werness, BA; Dyson, N; Phelps, WC; Harlow, E; Howley, PM. Complex formation of human papillomavirus E7 proteins with the retinoblastoma tumor suppressor gene product. *The EMBO journal*, 1989, Dec 20, 8(13), 4099-105.

[98] Imai, Y; Matsushima, Y; Sugimura, T; Terada, M. Purification and characterization of human papillomavirus type 16 E7 protein with preferential binding capacity to the underphosphorylated form of retinoblastoma gene product. *Journal of virology*, 1991, Sep, 65(9), 4966-72.

[99] Martin, LG; Demers, GW; Galloway, DA. Disruption of the G1/S transition in human papillomavirus type 16 E7-expressing human cells is associated with altered regulation of cyclin E. *Journal of virology*, 1998, Feb, 72(2), 975-85.

[100] Schulze, A; Mannhardt, B; Zerfass-Thome, K; Zwerschke, W; Jansen-Durr, P. Anchorage-independent transcription of the cyclin A gene induced by the E7

oncoprotein of human papillomavirus type 16. *Journal of virology*, 1998, Mar, 72(3), 2323-34.

[101] Hwang, SG; Lee, D; Kim, J; Seo, T; Choe, J. Human papillomavirus type 16 E7 binds to E2F1 and activates E2F1-driven transcription in a retinoblastoma protein-independent manner. *The Journal of biological chemistry*, 2002, Jan 25, 277(4), 2923-30.

[102] Lukas, J; Parry, D; Aagaard, L; Mann, DJ; Bartkova, J; Strauss, M; et al. Retinoblastoma-protein-dependent cell-cycle inhibition by the tumour suppressor p16. *Nature*, 1995, Jun 8, 375(6531), 503-6.

[103] Mann, DJ; Jones, NC. E2F-1 but not E2F-4 can overcome p16-induced G1 cell-cycle arrest. *Curr.Biol.*, 1996, Apr 1, 6(4), 474-83.

[104] Giarre, M; Caldeira, S; Malanchi, I; Ciccolini, F; Leao, MJ; Tommasino, M. Induction of pRb degradation by the human papillomavirus type 16 E7 protein is essential to efficiently overcome p16INK4a-imposed G1 cell cycle Arrest. *J. Virol*, 2001, May, 75(10), 4705-12.

[105] Sano, T; Oyama, T; Kashiwabara, K; Fukuda, T; Nakajima, T. Expression status of p16 protein is associated with human papillomavirus oncogenic potential in cervical and genital lesions. T*he American journal of pathology*, 1998, Dec, 153(6), 1741-8.

[106] Menges, CW; Baglia, LA; Lapoint, R; McCance, DJ. Human papillomavirus type 16 E7 up-regulates AKT activity through the retinoblastoma protein. *Cancer research*, 2006, Jun 1, 66(11), 5555-9.

[107] Charette, ST; McCance, DJ. The E7 protein from human papillomavirus type 16 enhances keratinocyte migration in an Akt-dependent manner. *Oncogene.* 2007, May 28.

[108] Khwaja, A; Rodriguez-Viciana, P; Wennstrom, S; Warne, PH; Downward, J. Matrix adhesion and Ras transformation both activate a phosphoinositide 3-OH kinase and protein kinase B/Akt cellular survival pathway. *The EMBO journal*, 1997, May 15, 16(10), 2783-93.

[109] Rangarajan, A; Syal, R; Selvarajah, S; Chakrabarti, O; Sarin, A; Krishna, S. Activated Notch1 signaling cooperates with papillomavirus oncogenes in transformation and generates resistance to apoptosis on matrix withdrawal through PKB/Akt. *Virology*, 2001, Jul 20, 286(1), 23-30.

[110] DeCaprio, JA; Ludlow, JW; Figge, J; Shew, JY; Huang, CM; Lee, WH; et al. SV40 large tumor antigen forms a specific complex with the product of the retinoblastoma susceptibility gene. *Cell*, 1988, Jul, 15, 54(2), 275-83.

[111] Chen S; Paucha, E. Identification of a region of simian virus 40 large T antigen required for cell transformation. *Journal of virology*, 1990, Jul, 64(7), 3350-7.

[112] Hiebert, SW; Chellappan, SP; Horowitz, JM; Nevins, JR. The interaction of RB with E2F coincides with an inhibition of the transcriptional activity of E2F. *Genes and development*, 1992, Feb, 6(2), 177-85.

[113] Qin, XQ; Chittenden, T; Livingston, DM; Kaelin, WG; Jr. Identification of a growth suppression domain within the retinoblastoma gene product. *Genes and development*, 1992, Jun, 6(6), 953-64.

[114] Lee, JO; Russo, AA; Pavletich, NP. Structure of the retinoblastoma tumour-suppressor pocket domain bound to a peptide from HPV E7. *Nature*, 1998, Feb 26, 391(6670), 859-65.

[115] Patrick, DR; Oliff, A; Heimbrook, DC. Identification of a novel retinoblastoma gene product binding site on human papillomavirus type 16 E7 protein. *The Journal of biological chemistry*, 1994, Mar 4, 269(9), 6842-50.

[116] Xiao, B; Spencer, J; Clements, A; Ali-Khan, N; Mittnacht, S; Broceno, C; et al. Crystal structure of the retinoblastoma tumor suppressor protein bound to E2F and the molecular basis of its regulation. *Proceedings of the National Academy of Sciences of the United States of America*, 2003, Mar 4, 100(5), 2363-8.

[117] Liu, X; Clements, A; Zhao, K; Marmorstein, R. Structure of the human Papillomavirus E7 oncoprotein and its mechanism for inactivation of the retinoblastoma tumor suppressor. *The Journal of biological chemistry*, 2006, Jan 6, 281(1), 578-86.

[118] Jones, DL; Munger, K. Analysis of the p53-mediated G1 growth arrest pathway in cells expressing the human papillomavirus type 16 E7 oncoprotein. *Journal of virology*, 1997, Apr, 71(4), 2905-12.

[119] Ciccolini, F; Di Pasquale, G; Carlotti, F; Crawford, L; Tommasino, M. Functional studies of E7 proteins from different HPV types. *Oncogene*, 1994, Sep, 9(9), 2633-8.

[120] Schmitt, A; Harry, JB; Rapp, B; Wettstein, FO; Iftner, T. Comparison of the properties of the E6 and E7 genes of low- and high-risk cutaneous papillomaviruses reveals strongly transforming and high Rb-binding activity for the E7 protein of the low-risk human papillomavirus type 1. *Journal of virology*, 1994, Nov, 68(11), 7051-9.

[121] Davies, R; Hicks, R; Crook, T; Morris, J; Vousden, K. Human papillomavirus type 16 E7 associates with a histone H1 kinase and with p107 through sequences necessary for transformation. *Journal of virology*, 1993, May, 67(5), 2521-8.

[122] ,Classon M; Dyson, N. p107 and p130, versatile proteins with interesting pockets. *Experimental cell research*, 2001, Mar 10, 264(1), 135-47.

[123] Arroyo, M; ,Bagchi S; Raychaudhuri, P. Association of the human papillomavirus type 16 E7 protein with the S-phase-specific E2F-cyclin A complex. *Molecular and cellular biology*, 1993, Oct, 13(10), 6537-46.

[124] Lam, EW; Morris, JD; Davies, R; Crook, T; Watson, RJ; Vousden, KH. HPV16 E7 oncoprotein deregulates B-myb expression: correlation with targeting of p107/E2F complexes. *The EMBO journal*, 1994, Feb 15, 13(4), 871-8.

[125] Zerfass, K; Levy, LM; Cremonesi, C; Ciccolini, F; Jansen-Durr, P; Crawford, L; et al. Cell cycle-dependent disruption of E2F-p107 complexes by human papillomavirus type 16 E7. *The Journal of general virology*, 1995, Jul, 76 (Pt 7), 1815-20.

[126] Grana, X; Garriga, J; Mayol, X. Role of the retinoblastoma protein family, pRB, p107 and p130 in the negative control of cell growth. *Oncogene*, 1998, Dec 24, 17(25), 3365-83.

[127] Berezutskaya, E; Yu, B; Morozov, A; Raychaudhuri, P; Bagchi, S. Differential regulation of the pocket domains of the retinoblastoma family proteins by the HPV16 E7 oncoprotein. *Cell Growth Differ*, 1997, Dec, 8(12), 1277-86.

[128] Firzlaff, JM; ,Luscher B; Eisenman, RN. Negative charge at the casein kinase II phosphorylation site is important for transformation but not for Rb protein binding by the E7 protein of human papillomavirus type 16. *Proceedings of the National Academy of Sciences of the United States of America*, 1991 Jun 15, 88(12), 5187-91.

[129] Banks, L; Edmonds, C; Vousden, KH. Ability of the HPV16 E7 protein to bind RB and induce DNA synthesis is not sufficient for efficient transforming activity in NIH3T3 cells. *Oncogene*, 1990, Sep, 5(9), 1383-9.

[130] Phelps, WC; Munger, K; Yee, CL; Barnes, JA; Howley, PM. Structure-function analysis of the human papillomavirus type 16 E7 oncoprotein. *Journal of virology*, 1992, Apr, 66(4), 2418-27.

[131] Tommasino, M; Adamczewski, JP; Carlotti, F; Barth, CF; Manetti, R; Contorni, M; et al. HPV16 E7 protein associates with the protein kinase p33CDK2 and cyclin A. *Oncogene*, 1993, Jan, 8(1), 195-202.

[132] McIntyre, MC; Ruesch, MN; Laimins, LA. Human papillomavirus E7 oncoproteins bind a single form of cyclin E in a complex with cdk2 and p107. *Virology*, 1996, Jan 1, 215(1), 73-82.

[133] He, W; Staples, D; Smith, C; Fisher, C. Direct activation of cyclin-dependent kinase 2 by human papillomavirus E7. *Journal of virology*, 2003, Oct, 77(19), 10566-74.

[134] Ruesch, MN; Laimins, LA. Initiation of DNA synthesis by human papillomavirus E7 oncoproteins is resistant to p21-mediated inhibition of cyclin E-cdk2 activity. *Journal of virology*, 1997, Jul, 71(7), 5570-8.

[135] Jian, Y; Schmidt-Grimminger, DC; Chien, WM; Wu, X; Broker, TR; Chow, LT. Post-transcriptional induction of p21cip1 protein by human papillomavirus E7 inhibits unscheduled DNA synthesis reactivated in differentiated keratinocytes. *Oncogene*, 1998, Oct 22, 17(16), 2027-38.

[136] Noya, F; Chien, WM; Broker, TR; Chow, LT. p21cip1 Degradation in differentiated keratinocytes is abrogated by costabilization with cyclin E induced by human papillomavirus E7. *Journal of virology*, 2001, Jul, 75(13), 6121-34.

[137] Helt, AM; Funk, JO; Galloway, DA. Inactivation of both the retinoblastoma tumor suppressor and p21 by the human papillomavirus type 16 E7 oncoprotein is necessary to inhibit cell cycle arrest in human epithelial cells. *Journal of virology*, 2002, Oct, 76(20), 10559-68.

[138] Besson, A; Gurian-West, M; Schmidt, A; Hall, A; Roberts, JM. p27Kip1 modulates cell migration through the regulation of RhoA activation. *Genes and development*, 2004, Apr 15, 18(8), 862-76.

[139] Li, R; Wheeler, TM; Dai, H; Sayeeduddin, M; Scardino, PT; Frolov, A; et al. Biological correlates of p27 compartmental expression in prostate cancer. *The Journal of urology*, 2006, Feb, 175(2), 528-32.

[140] McAllister, SS; Becker-Hapak, M; Pintucci, G; Pagano, M; Dowdy, SF. Novel p27(kip1) C-terminal scatter domain mediates Rac-dependent cell migration independent of cell cycle arrest functions. *Molecular and cellular biology*, 2003, Jan, 23(1), 216-28.

[141] Boutros, R; Dozier, C; Ducommun, B. The when and wheres of CDC25 phosphatases. *Current opinion in cell biology*, 2006, Apr, 18(2), 185-91.

[142] Katich, SC; Zerfass-Thome, K; Hoffmann, I. Regulation of the Cdc25A gene by the human papillomavirus Type 16 E7 oncogene. *Oncogene*, 2001, Feb 1, 20(5), 543-50.

[143] Brehm, A; Miska, EA; McCance, DJ; Reid, JL; Bannister, AJ; Kouzarides, T. Retinoblastoma protein recruits histone deacetylase to repress transcription. *Nature*, 1998, Feb 5, 391(6667), 597-601.

[144] Luo, RX; Postigo, AA; Dean DC. Rb interacts with histone deacetylase to repress transcription. *Cell*, 1998, Feb 20, 92(4), 463-73.

[145] Harbour, JW; Dean, DC. The Rb/E2F pathway: expanding roles and emerging paradigms. *Genes and development*, 2000, Oct 1, 14(19), 2393-409.

[146] Dick, FA; Sailhamer, E; Dyson, NJ. Mutagenesis of the pRB pocket reveals that cell cycle arrest functions are separable from binding to viral oncoproteins. *Molecular and cellular biology*, 2000, May, 20(10), 3715-27.

[147] Li, H; Ou, X; Xiong, J; Wang, T. HPV16E7 mediates HADC chromatin repression and downregulation of MHC class I genes in HPV16 tumorigenic cells through interaction with an MHC class I promoter. *Biochemical and biophysical research communications*, 2006, Nov 3, 349(4), 1315-21.

[148] Huh, KW; DeMasi, J; Ogawa, H; Nakatani, Y; Howley, PM; Munger, K. Association of the human papillomavirus type 16 E7 oncoprotein with the 600-kDa retinoblastoma protein-associated factor, p600. *Proceedings of the National Academy of Sciences of the United States of America*, 2005, Aug 9, 102(32), 11492-7.

[149] Maldonado, E; Cabrejos, ME; Banks, L; Allende, JE. Human papillomavirus-16 E7 protein inhibits the DNA interaction of the TATA binding transcription factor. *Journal of cellular biochemistry*, 2002, 85(4), 663-9.

[150] Bernat, A; Avvakumov, N; Mymryk, JS; Banks, L. Interaction between the HPV E7 oncoprotein and the transcriptional coactivator p300. *Oncogene*, 2003, Sep 11, 22(39), 7871-81.

[151] Huang, SM; McCance, DJ. Down regulation of the interleukin-8 promoter by human papillomavirus type 16 E6 and E7 through effects on CREB binding protein/p300 and P/CAF. *Journal of virology*, 2002, Sep, 76(17), 8710-21.

[152] Turnell, AS; Mymryk, JS. Roles for the coactivators CBP and p300 and the APC/C E3 ubiquitin ligase in E1A-dependent cell transformation. *British journal of cancer*, 2006, Sep 4, 95(5), 555-60.

[153] Patel, D; Huang, SM; Baglia, LA; McCance, DJ. The E6 protein of human papillomavirus type 16 binds to and inhibits co-activation by CBP and p300. *The EMBO journal*, 1999, Sep 15, 18(18), 5061-72.

[154] Bernat, A; Massimi, P; Banks, L. Complementation of a p300/CBP defective-binding mutant of adenovirus E1a by human papillomavirus E6 proteins. *The Journal of general virology*, 2002, Apr, 83(Pt 4), 829-33.

[155] Vousden, KH; Vojtesek, B; Fisher, C; Lane, D. HPV-16 E7 or adenovirus E1A can overcome the growth arrest of cells immortalized with a temperature-sensitive p53. *Oncogene*, 1993, Jun, 8(6), 1697-702.

[156] Demers, GW; Foster, SA; Halbert, CL; Galloway, DA. Growth arrest by induction of p53 in DNA damaged keratinocytes is bypassed by human papillomavirus 16 E7. *Proceedings of the National Academy of Sciences of the United States of America.* 1994, May 10, 91(10), 4382-6.

[157] Demers, GW; Halbert, CL; Galloway, DA. Elevated wild-type p53 protein levels in human epithelial cell lines immortalized by the human papillomavirus type 16 E7 gene. *Virology*, 1994, Jan, 198(1), 169-74.

[158] Seavey, SE; Holubar, M; Saucedo, LJ; Perry, ME. The E7 oncoprotein of human papillomavirus type 16 stabilizes p53 through a mechanism independent of p19(ARF). *Journal of virology*, 1999, Sep, 73(9), 7590-8.

[159] Pan, H; Griep, AE. Altered cell cycle regulation in the lens of HPV-16 E6 or E7 transgenic mice: implications for tumor suppressor gene function in development. *Genes and development*, 1994, Jun 1, 8(11), 1285-99.

[160] Thomas, M; Banks, L. Inhibition of Bak-induced apoptosis by HPV-18 E6. *Oncogene*, 1998, Dec 10, 17(23), 2943-54.
[161] Massimi, P; Banks, L. Repression of p53 transcriptional activity by the HPV E7 proteins. *Virology*, 1997, Jan 6, 227(1), 255-9.
[162] Stoppler, H; Stoppler, MC; Johnson, E; Simbulan-Rosenthal, CM; Smulson, ME; Iyer, S; et al. The E7 protein of human papillomavirus type 16 sensitizes primary human keratinocytes to apoptosis. *Oncogene*, 1998, Sep 10, 17(10), 1207-14.
[163] Thyrell, L; Sangfelt, O; Zhivotovsky, B; Pokrovskaja, K; Wang, Y; Einhorn, S; et al. The HPV-16 E7 oncogene sensitizes malignant cells to IFN-alpha-induced apoptosis. *J Interferon Cytokine Res.*, 2005, Feb, 25(2), 63-72.
[164] Yuan, H; Fu, F; Zhuo, J; Wang, W; Nishitani, J; An, DS; et al. Human papillomavirus type 16 E6 and E7 oncoproteins upregulate c-IAP2 gene expression and confer resistance to apoptosis. *Oncogene*, 2005, Jul 28, 24(32), 5069-78.
[165] Thompson, DA; Zacny, V; Belinsky, GS; Classon, M; Jones, DL; Schlegel, R; et al. The HPV E7 oncoprotein inhibits tumor necrosis factor alpha-mediated apoptosis in normal human fibroblasts. *Oncogene*, 2001, Jun 21, 20(28), 3629-40.
[166] Duensing, S; Lee, LY; Duensing, A; Basile, J; Piboonniyom, S; Gonzalez, S; et al. The human papillomavirus type 16 E6 and E7 oncoproteins cooperate to induce mitotic defects and genomic instability by uncoupling centrosome duplication from the cell division cycle. *Proceedings of the National Academy of Sciences of the United States of America*, 2000, Aug 29, 97(18), 10002-7.
[167] Hashida, T; Yasumoto, S. Induction of chromosome abnormalities in mouse and human epidermal keratinocytes by the human papillomavirus type 16 E7 oncogene. *The Journal of general virology*, 1991, Jul, 72 (Pt 7), 1569-77.
[168] Duensing, S; Munger, K. Human papillomavirus type 16 E7 oncoprotein can induce abnormal centrosome duplication through a mechanism independent of inactivation of retinoblastoma protein family members. *Journal of virology*, 2003, Nov, 77(22), 12331-5.
[169] Mantel, C; Braun, SE; Reid, S; Henegariu, O; Liu, L; Hangoc, G; et al. p21(cip-1/waf-1) deficiency causes deformed nuclear architecture, centriole overduplication, polyploidy, and relaxed microtubule damage checkpoints in human hematopoietic cells. *Blood*, 1999, Feb 15, 93(4), 1390-8.
[170] Lacey, KR; Jackson, PK; Stearns, T. Cyclin-dependent kinase control of centrosome duplication. *Proceedings of the National Academy of Sciences of the United States of America*, 1999, Mar 16, 96(6), 2817-22.
[171] Duensing, A; Liu, Y; Perdreau, SA; Kleylein-Sohn, J; Nigg, EA; Duensing, S. Centriole overduplication through the concurrent formation of multiple daughter centrioles at single maternal templates. *Oncogene*, 2007, Apr 16.
[172] Nguyen, CL; Munger, K. The Human Papillomavirus Type 16 E7 Oncoprotein Associates with the Centrosomal Component {gamma}-Tubulin. *Journal of virology*, 2007, Oct 3.
[173] Duensing, S; Munger, K. The human papillomavirus type 16 E6 and E7 oncoproteins independently induce numerical and structural chromosome instability. *Cancer research*, 2002, Dec 1, 62(23), 7075-82.
[174] Koopman, LA; Szuhai, K; van Eendenburg, JD; Bezrookove, V; Kenter, GG; Schuuring, E; et al. Recurrent integration of human papillomaviruses 16, 45, and 67

near translocation breakpoints in new cervical cancer cell lines. *Cancer research*, 1999, Nov 1, 59(21), 5615-24.

[175] Song, S; Gulliver, GA; Lambert, PF. Human papillomavirus type 16 E6 and E7 oncogenes abrogate radiation-induced DNA damage responses in vivo through p53-dependent and p53-independent pathways. *Proceedings of the National Academy of Sciences of the United States of America*, 1998, Mar 3, 95(5), 2290-5.

[176] Hasan, UA; Caux, C; Perrot, I; Doffin, AC; Menetrier-Caux, C; Trinchieri, G; et al. Cell proliferation and survival induced by Toll-like receptors is antagonized by type I IFNs. *Proceedings of the National Academy of Sciences of the United States of America*, 2007, May 8, 104(19), 8047-52.

[177] Tanaka, N; Ishihara, M; Lamphier, MS; Nozawa, H; Matsuyama, T; Mak, TW; et al. Cooperation of the tumour suppressors IRF-1 and p53 in response to DNA damage. *Nature*, 1996, Aug 29, 382(6594), 816-8.

[178] Georgopoulos, NT; Proffitt, JL; Blair, GE. Transcriptional regulation of the major histocompatibility complex (MHC) class I heavy chain, TAP1 and LMP2 genes by the human papillomavirus (HPV) type 6b, 16 and 18 E7 oncoproteins. *Oncogene*, 2000, Oct 5, 19(42), 4930-5.

[179] Cromme, FV; Meijer, CJ; Snijders, PJ; Uyterlinde, A; Kenemans, P; Helmerhorst, T; et al. Analysis of MHC class I and II expression in relation to presence of HPV genotypes in premalignant and malignant cervical lesions. *British journal of cancer*, 1993, Jun, 67(6), 1372-80.

[180] Cromme, FV; Airey, J; Heemels, MT; Ploegh, HL; Keating, PJ; Stern, PL; et al. Loss of transporter protein, encoded by the TAP-1 gene, is highly correlated with loss of HLA expression in cervical carcinomas. *The Journal of experimental medicine*, 1994, Jan 1, 179(1), 335-40.

[181] Bottley, G; Watherston, OG; Hiew, YL; Norrild, B; Cook, GP; Blair, GE. High-risk human papillomavirus E7 expression reduces cell-surface MHC class I molecules and increases susceptibility to natural killer cells. *Oncogene*, 2007, Sep 10.

[182] Frazer, IH; Thomas, R; Zhou, J; Leggatt, GR; Dunn, L; McMillan, N; et al. Potential strategies utilised by papillomavirus to evade host immunity. *Immunological reviews*, 1999, Apr, 168, 131-42.

[183] zur Hausen, H. Papillomaviruses in human cancers. *Proceedings of the Association of American Physicians*, 1999, Nov-Dec, 111(6), 581-7.

[184] Wright, TC; Jr. Cervical cancer screening in the 21st century: is it time to retire the PAP smear? *Clinical obstetrics and gynecology*, 2007, Jun, 50(2), 313-23.

[185] Harper, DM; Franco, EL; Wheeler, CM; Moscicki, AB; Romanowski, B; Roteli-Martins, CM; et al. Sustained efficacy up to 4.5 years of a bivalent L1 virus-like particle vaccine against human papillomavirus types 16 and 18: follow-up from a randomised control trial. *Lancet*, 2006, Apr 15, 367(9518), 1247-55.

[186] Koutsky, LA; Ault, KA; Wheeler, CM; Brown, DR; Barr, E; Alvarez, FB; et al. A controlled trial of a human papillomavirus type 16 vaccine. *The New England journal of medicine*, 2002, Nov 21, 347(21), 1645-51.

[187] Moore, DH. Chemotherapy for recurrent cervical carcinoma. *Current opinion in oncology*, 2006, Sep, 18(5), 516-9.

[188] Gammoh, N; Grm, HS; Massimi, P; Banks, L. Regulation of human papillomavirus type 16 E7 activity through direct protein interaction with the E2 transcriptional activator. *Journal of virology*, 2006, Feb, 80(4), 1787-97.

[189] Goodwin, EC; Yang, E; Lee, CJ; Lee, HW; DiMaio, D; Hwang, ES. Rapid induction of senescence in human cervical carcinoma cells. *Proceedings of the National Academy of Sciences of the United States of America*, 2000, Sep 26, 97(20), 10978-83.

[190] Lee, CJ; Suh, EJ; Kang, HT; Im, JS; Um, SJ; Park, JS; et al. Induction of senescence-like state and suppression of telomerase activity through inhibition of HPV E6/E7 gene expression in cells immortalized by HPV16 DNA. *Experimental cell research*, 2002, Jul 15, 277(2), 173-82.

[191] Desaintes, C; Demeret, C; Goyat, S; Yaniv, M; Thierry, F. Expression of the papillomavirus E2 protein in HeLa cells leads to apoptosis. *The EMBO journal*, 1997, Feb 3, 16(3), 504-14.

[192] DeFilippis, RA; Goodwin, EC; Wu, L; DiMaio, D. Endogenous human papillomavirus E6 and E7 proteins differentially regulate proliferation, senescence, and apoptosis in HeLa cervical carcinoma cells. *Journal of virology*, 2003, Jan, 77(2), 1551-63.

[193] Wu, S; Meng, L; Wang, S; Wang, W; Xi, L; Tian, X; et al. Reversal of the malignant phenotype of cervical cancer CaSki cells through adeno-associated virus-mediated delivery of HPV16 E7 antisense RNA. *Clin. Cancer Res.*, 2006, Apr 1, 12(7 Pt 1), 2032-7.

[194] Hamada, K; Shirakawa, T; Gotoh, A; Roth, JA; Follen, M. Adenovirus-mediated transfer of human papillomavirus 16 E6/E7 antisense RNA and induction of apoptosis in cervical cancer. *Gynecologic oncology*, 2006, Dec, 103(3), 820-30.

[195] Kari, I; Syrjanen, S; Johansson, B; Peri, P; He, B; Roizman, B; et al. Antisense RNA directed to the human papillomavirus type 16 E7 mRNA from herpes simplex virus type 1 derived vectors is expressed in CaSki cells and downregulates E7 mRNA. *Virology journal*, 2007, 4, 47.

[196] Choo, CK; Ling, MT; Suen, CK; Chan, KW; Kwong, YL. Retrovirus-mediated delivery of HPV16 E7 antisense RNA inhibited tumorigenicity of CaSki cells. *Gynecologic oncology*, 2000, Sep, 78(3 Pt 1), 293-301.

[197] Sima, N; Wang, S; Wang, W; Kong, D; Xu, Q; Tian, X; et al. Antisense targeting human papillomavirus type 16 E6 and E7 genes contributes to apoptosis and senescence in SiHa cervical carcinoma cells. *Gynecologic oncology*, 2007, Aug, 106(2), 299-304.

[198] Jiang, M; Milner, J. Selective silencing of viral gene expression in HPV-positive human cervical carcinoma cells treated with siRNA, a primer of RNA interference. *Oncogene*, 2002, Sep 5, 21(39), 6041-8.

[199] Hall, AH; Alexander, KA. RNA interference of human papillomavirus type 18 E6 and E7 induces senescence in HeLa cells. *Journal of virology*, 2003, May, 77(10), 6066-9.

[200] Fujii, T; Saito, M; Iwasaki, E; Ochiya, T; Takei, Y; Hayashi, S; et al. Intratumor injection of small interfering RNA-targeting human papillomavirus 18 E6 and E7 successfully inhibits the growth of cervical cancer. *International journal of oncology*, 2006, Sep, 29(3), 541-8.

[201] Tang, S; Tao, M; McCoy, JP; Jr., Zheng, ZM. Short-term induction and long-term suppression of HPV16 oncogene silencing by RNA interference in cervical cancer cells. *Oncogene*, 2006, Mar 30, 25(14), 2094-104.

[202] Gu, W; Putral, L; Hengst, K; Minto, K; Saunders, NA; Leggatt, G; et al. Inhibition of cervical cancer cell growth in vitro and in vivo with lentiviral-vector delivered short hairpin *RNA targeting human papillomavirus E6 and E7 oncogenes. Cancer gene therapy*, 2006, Nov, 13(11), 1023-32.

[203] Lu, D; Chatterjee, S; Brar, D; Wong, KK. Jr. Ribozyme-mediated in vitro cleavage of transcripts arising from the major transforming genes of human papillomavirus type 16. *Cancer gene therapy*, 1994, Dec, 1(4), 267-77.

[204] Chen, Z; Kamath, P; Zhang, S; Weil, MM; Shillitoe, EJ. Effectiveness of three ribozymes for cleavage of an RNA transcript from human papillomavirus type 18. *Cancer gene therapy*, 1995, Dec, 2(4), 263-71.

[205] Alvarez-Salas, LM; Cullinan, AE; Siwkowski, A; Hampel, A; DiPaolo, JA. Inhibition of HPV-16 E6/E7 immortalization of normal keratinocytes by hairpin ribozymes. *Proceedings of the National Academy of Sciences of the United States of America*, 1998, Feb 3, 95(3), 1189-94.

[206] Zheng, Y; Zhang, J; Rao, Z. Ribozyme targeting HPV16 E6E7 transcripts in cervical cancer cells suppresses cell growth and sensitizes cells to chemotherapy and radiotherapy. *Cancer biology and therapy*, 2004, Nov, 3(11), 1129-34, discussion 35-6.

[207] Kunke, D; Grimm, D; Denger, S; Kreuzer, J; Delius, H; Komitowski, D; et al. Preclinical study on gene therapy of cervical carcinoma using adeno-associated virus vectors. *Cancer gene therapy*, 2000, May, 7(5), 766-77.

[208] Nauenburg, S; Zwerschke, W; Jansen-Durr, P. Induction of apoptosis in cervical carcinoma cells by peptide aptamers that bind to the HPV-16 E7 oncoprotein. *Faseb. J*, 2001, Mar, 15(3), 592-4.

[209] Accardi, L; Dona, MG; Di Bonito, P; Giorgi, C. Intracellular anti-E7 human antibodies in single-chain format inhibit proliferation of HPV16-positive cervical carcinoma cells. *International journal of cancer*, 2005, Sep 10, 116(4), 564-70.

[210] Griffin, H; Elston, R; Jackson, D; Ansell, K; Coleman, M; Winter, G; et al. Inhibition of papillomavirus protein function in cervical cancer cells by intrabody targeting. *Journal of molecular biology*, 2006, Jan 20, 355(3), 360-78.

[211] Karl, T; Seibert, N; Stohr, M; Osswald, H; Rosl, F; Finzer, P. Sulindac induces specific degradation of the HPV oncoprotein E7 and causes growth arrest and apoptosis in cervical carcinoma cells. *Cancer letters*, 2007, Jan 8, 245(1-2), 103-11.

[212] Andrei, G; Snoeck, R; Schols, D; De Clercq, E. Induction of apoptosis by cidofovir in human papillomavirus (HPV)-positive cells. *Oncology research*, 2000, 12(9-10), 397-408.

[213] Hostetler, KY; Rought, S; Aldern, KA; Trahan, J; Beadle, JR; Corbeil, J. Enhanced antiproliferative effects of alkoxyalkyl esters of cidofovir in human cervical cancer cells in vitro. *Molecular cancer therapeutics*, 2006, Jan, 5(1), 156-9.

[214] Divya, CS; Pillai, MR. Antitumor action of curcumin in human papillomavirus associated cells involves downregulation of viral oncogenes, prevention of NFkB and AP-1 translocation, and modulation of apoptosis. *Molecular carcinogenesis*, 2006, May, 45(5), 320-32.

[215] Yim, EK; Lee, MJ; Lee, KH; Um, SJ; Park, JS. Antiproliferative and antiviral mechanisms of ursolic acid and dexamethasone in cervical carcinoma cell lines. *Int. J. Gynecol. Cancer*, 2006, Nov-Dec, 16(6), 2023-31.

[216] Beerheide, W; Bernard, HU; Tan, YJ; Ganesan, A; Rice, WG; Ting, AE. Potential drugs against cervical cancer: zinc-ejecting inhibitors of the human papillomavirus type 16 E6 oncoprotein. *Journal of the National Cancer Institute*, 1999, Jul 21, 91(14), 1211-20.

[217] Beerheide, W; Sim, MM; Tan, YJ; Bernard, HU; Ting, AE. Inactivation of the human papillomavirus-16 E6 oncoprotein by organic disulfides. *Bioorganic and medicinal chemistry*, 2000, Nov, 8(11), 2549-60.

[218] Rogers, JP; Beuscher, AEt; Flajolet, M; McAvoy, T; Nairn, AC; Olson, AJ; et al. Discovery of protein phosphatase 2C inhibitors by virtual screening. *Journal of medicinal chemistry*, 2006, Mar 9, 49(5), 1658-67.

[219] Lu, Y; Nikolovska-Coleska, Z; Fang, X; Gao, W; Shangary, S; Qiu, S; et al. Discovery of a nanomolar inhibitor of the human murine double minute 2 (*MDM2)-p53 interaction through an integrated, virtual database screening strategy. Journal of medicinal chemistry*, 2006 Jun 29, 49(13), 3759-62.

[220] Nassar, N; Cancelas, J; Zheng, J; Williams, DA; Zheng, Y. Structure-function based design of small molecule inhibitors targeting Rho family GTPases. *Current topics in medicinal chemistry*, 2006, 6(11), 1109-16.

[221] Trosset, JY; Dalvit, C; Knapp, S; Fasolini, M; Veronesi, M; Mantegani, S; et al. Inhibition of protein-protein interactions: the discovery of druglike beta-catenin inhibitors by combining virtual and biophysical screening. *Proteins*, 2006, Jul 1, 64(1), 60-7.

[222] Song, H; Wang, R; Wang, S; Lin, J. A low-molecular-weight compound discovered through virtual database screening inhibits Stat3 function in breast cancer cells. *Proceedings of the National Academy of Sciences of the United States of America*, 2005, Mar 29, 102(13), 4700-5.

[223] Cavasotto, CN; Ortiz, MA; Abagyan, RA; Piedrafita, FJ. In silico identification of novel EGFR inhibitors with antiproliferative activity against cancer cells. *Bioorganic and medicinal chemistry letters*, 2006 Apr 1, 16(7), 1969-74.

In: Oncoproteins: Types and Detection
Editor: Jeremy R. Davis

ISBN: 978-1-61761-551-1

Chapter 9

Oncogene Proteins: New Research Oncogene Proteins as Tumor Markers

Viroj Wiwanitkit
Department of Laboratory Medicine, Faculty of Medicine,
Chulalongkorn University, Bangkok, Thailand

Abstract

Tumor markersis a group of important laboratory diagnostic tools. In this review, details on the general concepts of tumor markers their usefullness and limitations are collected and presented. Basically, a specific tumor marker is a fusion protein associated with a malignant process in which an oncogene is translocated. Oncogene proteins can serve as tumor markers. Here, important oncogene proteins as serum and tissue tumor markers, will be presented and discussed. In addition, the clinical application of oncogene proteins as tumor markers for important tumors in medicine will be mentioned.

Introduction to Tumor Markers

Cancer is an important pathological condition. The diagnosis of cancer is hard. Without laboratory investigation, cancer is usually detected by chance due to the history taking and physical examination. The mass is the common clinical manifestation. For cancer clinical pathology, the tool for diagnosis of cancer is focused for a long time. Tumor marker is a group of laboratory investigations. Molecules occurring in the human body, associated with cancer are tumor markers. Tumor marker can be used for diagnosis or management of disease. An ideal marker is a "blood test" that is positive only when the patient has the disease. In clinical pathology, tumor markers have a very long history. The first tumor marker is the Bence Jones protein, a urine protein that was first described by Bence Jones in 1847 [1].

This protein is the tumor marker for multiple myeloma, a cancer of the plasma cell. The other classical tumor markers include acid phosphatase in advanced prostate cancer which was discovered by Gutman and Gutman in 1938 [2]. Oncofetal antigens and prostate specific antigen are examples of new tumor markers discovered in late twentieth century. Five groups of serum tumor markers are available at present [3]. The enzyme and hormone markers are not specific since these two groups can be secreted from the normal cells. Of five groups, tumor antigens are the most specific. Two subgroups of tumor antigens, oncofetal and non-oncofetal proteins are mentioned [3].

Basically, protein synthesis starts within the nucleus. DNA translates via mRNA into protein in cytosol and secretes out to the membrane and is transported through cell membranes into the external environment. These proteins can then be detected in serum. Any abnormal proteins due to the carcinogenesis process can be a candidate for the tumor marker test. Monoclonal antibodies can be developed and used to detect these proteins on the cancer cell surface or in the serum. The five main advantages of tumor markers includes screening, diagnosis, monitoring therapy, monitoring remission and following-up (Table 2).

Table 1. Examples of tumor markers

Types	Example	Diagnosis usefulness
Enzymes	Acid Phosphatase	Leukemia, cholanigiocarcinoma
	Neuron specific enolase	Neurological malignancy
Hormones	HCG	Choriocarcinoma
	ADH	Lung cancer
	Calcitonin	Thyroid C cell cancer
	Parathormone	Parathyroid cancer
Tumor antigens	CEA	Colon cancer
	CA 125	Ovarian cancer
	CA 19-9	Gastrointestinal cancer
	PSA	Prostate cancer
	AFP	Liver cancer
Immunoglobulins	Immunoglobulin	Plasmacytoma
Other	Bence Jones Protein	Multiple myeloma

Table 2. Examples of advantage of some tumor markers

Advantages	Example	Diagnosis usefulness
screening	PSA	Prostate cancer
diagnosis	Bence Jones Protein	Multiple myeloma
monitoring therapy	AFP	Liver cancer
monitoring remission	Calcitonin	Thyroid C cell cancer
following-up	HCG	Choriocarcinoma

Some Problems on the Present Tumor Marker Testing

1. Lack of Good Test

Screening and diagnosis are both focused properties of good tumor markers. For screening, high sensitivity is needed. While for diagnosis, high specificity is needed. Any test with these two combined properties are difficult to find. Of several tumor markers, PSA is the best marker in clinical pathology [4].

Most of the present tumor markers lack specificity due to the high false negatives. Benign diseases positive CA 125 [5] or CEA [6] are common. Some physiological underlying can affect the analytical result such as smokers having a raised CEA [7]. Also, cancer heterogeneity usually exists for many organs [8]. In addition, many patients die with, not from, cancers.

2. How Can We Manage the Tumor Marker Test?

Due to the preventive medicine principle, early diagnosis is good preventive action. Screening is a widely used tumor marker. However, there are many problems such as, "Is a negative reassuring?" and "What does a positive indicate?". Management of anxiety due to the test result is a necessary point [9 - 10]. Sometimes, the increased level of tumor markers can be seen in benign diseases. On the other hand, some increase can be detected in the late stage, unresectable mass. These facts are the flaws of tumor marker screening.

For following up, a decrease after therapy usually means response, however, "normal" does not mean remission [8]. The query in palliative therapy is treating the symptom or the tumor marker level. Using the laboratory results accompanied with the signs and symptoms is suggested. However, it should be noted that not all symptoms are cancer progression. At the end of therapy, the physician should beware of misleading. A normal level does not mean cured, does not mean complete remission, may not even mean improvement. This can create anxiety, false reassurance and false hope [8].

3. Decision Making due to the Tumor Marker Investigation Result

Decision making due to the tumor marker investigation result is sometimes very hard since tumor marker results sometimes seems aberrant. Some common aberrant results include increased to two markers. Indeed, the possible clues for this phenomenon are 1) there are two existing growing tumors within the patient, which is rare but possible, 2) there is something wrong in the investigation and 3) there is possibly asecond tumor on top of the primary lesion. In addition, it should be noted that the tumor marker sometimes increases in the improper time such as increase in the benign disease due to the interference of external variation especially for cigarette smoking, which can bring to the over therapeutic options. On the other hand, it sometimes can be seen that the tumor marker just rise when the disease is very advanced and untreatable, resection is not possible. These facts bring the conclusion to the physicians that they should rely on the tumor marker accompanied with the patients' signs and symptoms in making a decision.

4. Communication Gaps between the Patients and Physician

"Does the patient understand?" is the main question when a physician in charge ordering the tumor marker testing. Explanation is necessary. Creating of anxiety have to be managed. For the oncologist, interpretation of the test result should be carefully performed. Sometimes, multiple markers elevated and sometimes marker falling, mass increasing can be seen. In addition, the oncologist should be accessible to the family. Physicians who knows the patient best, must answer the questions "why?" or "what next"? from the family. The special technique to tell the bad news to the patients and his/her family must be used.

5. Availability of the Tumor Marker Testing

The tumor marker testing is considered a special test in laboratory test in medicine. The availability of the tumor marker testing is a concern. It cannot be denied that there is no equilibrium in health service assessment in different settings. Rural and remote setting usually lack for many medical facilities including the laboratory investigation. Tumor marker testing is usually lack in those distance settings. However, the good referring system and good clinical skill in decision making for request the test in the limited resource setting may be the clue for resolving of the problem of this disequilibrium.

Oncogene Proteins as Tumor Marker Candidates

Oncogenesis is a rather complicated process. As previously mentioned, the underlying genetic abnormalities accompanied with the external environmental insults are the main contributors to cancer development. Recent advances brought the identification of cellular genes which are involved in the starting and progression of tumorigenesis. Several types of cancers have been shown to be derived from cells that have extensively mutated both alleles of one or both of these genes, resulting in a loss-of-function mutation [11 – 13]. The proto-oncogenes, which normally participate in the regulation of cell proliferation and differentiation, can become oncogenes through alterations in the regulation of their expression and/or their coding sequences [11]. The proto-oncogenes can be grouped into different categories based on their protein products, i.e. protein kinases, growth factors, growth factor receptors, and DNA binding proteins. The analysis of the molecular mechanisms governing multistep carcinogenesis became experimentally approachable since the identification and characterization in tumor cells of altered or activated versions of cellular oncogenes that normally control cell growth and differentiation [12]. The activating mutations confer new properties to the oncogene products and should therefore be considered as gain of function mutations [12]. The anti-oncogenes or tumor suppressor genes or recessive oncogenes are normally implicated in a negative regulation of cellular proliferation [11]. The suppression or reversion of the malignant phenotype by the introduction of a normal chromosome into a tumor cell line has lent support to the idea that a family of cellular genes are coding for factors capable to interact with the cell-growth control machinery [12]. These genes seem to reconstitute the normal control of cell growth even in the presence of an activated oncogene [12]. Anderson and Spandidos noted for evidences for the existence of tumor suppressor genes based on the following points: a) recessive cancer genes in higher and lower eukaryotes have been detected, b) studies on cell hybrids have implicated genes which suppress various stages in the malignant conversion of normal cells, c) the isolation from virally induced transformants of flat, non-tumorigenic revertants in which expression of the transforming gene is not down-regulated suggests the presence of genes which suppress the effects of transformation. and d) blocks to differentiation can be bypassed by inducing compounds or differentiation factors [14]. The contribution of oncogene to the tumorigenic phenotype is dominant or positive regulation since they act also in the presence of the homologous wild-type allele while the loss of then tumor suppressor activity contributes to tumorigenesis in a recessive manner since the tumor phenotype appears when both alleles of an onco-suppressor gene are inactivated and the mutations affecting their normal functions belong to the type "loss of function" [11 - 12]. Balancing between oncogene and tumor suppressor gene lead to no tumorogenesis in normal condition. Common to most tumors, several regulatory circuits are altered during multistage tumor progression, most importantly, the control of proliferation, the balance between cell survival and programmed cell death (apoptosis), the communication with neighbouring cells and the extracellular matrix, the induction of tumor neovascularization and, finally, tumor cell migration, invasion and metastatic dissemination [15]. De-regulation of each of these processes represents a rate-limiting step for tumor development and, hence, has to be achieved by tumor cells in a highly selective manner during tumor progression [15].Apoptosis is a genetically controlled response by which eukaryotic cells undergo programmed cell death [16]. This phenomenon plays a major role in developmental pathways, provides a homeostatic balance of cell populations, and is deregulated in many diseases including cancer [16]. Elucidating the mechanisms that link cell cycle control with apoptosis will be of key importance in understanding tumor progression and designing new models of effective tumor therapy [16]. One may reasonably hope that defining a complete profile of alterations of both oncogene and tumor suppressor types of essential genes in the tumor or in the clinically healthy individual will be invaluable to assess prognosis or genetic predisposition, respectively [17].

At present, the main actions of proto-oncogenes during cellular proliferation are outlined and oncogenes are classified to their specific roles [18]. The relationships between proto-, viral-, and cellular oncogenes are analyzed and the several steps of the evolution of a p-onc towards a c-onc are well described. The oncogenes' products, their weight and cellular position are also presented [18]. The onco-suppressor genes, oncogenes' counterparts are also described, their localization being deduced from analysis of frequent breakpoints or mutation site/s in cancers [18]. Basically, in cancer cases, overaction of oncogene than suppressor genes can be expected. These genes are part of the normal complement of cells, and become altered in their structure or expression, during the development of the neoplastic phenotype [19]. Oncogene protein is an important product of oncogene that has many clinical implications.

Besides aiding in tumor diagnosis, the detection of such tumor-associated genes and their products allows the identification of individuals with an inherited predisposition to neoplastic growths, and the overexpression of many of these oncogene products has been shown to be a potential marker of tumor behavior and a predictor of treatment outcome and response [20]. The ability to utilize DNA and RNA probes for nucleic acid hybridization and polymerase chain reaction procedures in cell and tissue preparations of solid tumors and lymphoid proliferations expands and complements the information provided by immunohistochemical techniques [20].

Basically, a specific tumor marker is a fusion protein associated with a malignant process in which an oncogene is translocated. As previously mentioned, the main function of an oncogene is to control the growth of cells to make excessive proteins which further make the cells grow out of control and can result in cancer. Several oncogene proteins are mentioned in oncology. Those oncogene proteins can be served as tumor markers. Second generation tests could easily include PCR on mRNA, and/or in situ hybridization, both of which could be performed using blood samples [21]. Both methods would provide a faster means of testing a large number of cells, however, the methodologies must be improved through automation and computer-aided image analysis, respectively, in order to become useful routine tests [21]. Here, important oncogene proteins as serum and tissue tumor markers will be presented and discussed. In addition, the clinical application of oncogene proteins as tumor markers for important tumors in medicine will be mentioned.

1. Ras Oncogene Protein

The ras oncogene protein or P21 is an important oncogene protein. Point mutations in the cellular c- ras gene can also result in a mutant p21 protein that can transform human and other mammalian cells. Oncogene protein p21has been directly implicated in human tumors and might accounti for as much as 15-20% of all human tumors. One of the most commonly found transforming *ras* oncogenes in human tumours has a valine codon replacing the glycine codon at position 12 of the normal c-Ha-*ras* gene [22]. Tong et al found that one of the major differences between *ras* oncogene protein and normal protein is that the loop of the transforming ras protein that binds the β-phosphate of the guanine nucleotide is enlarged [22]. They also proposed that such a change in the "catalytic site' conformation could explain the reduced GTPase activity of the mutant,which keeps the protein in the GTP bound 'signal on9 state for a prolonged period of time [22]. The ras oncogene protein is known to occur frequently in human tumors, including those of occupational concern such as lung cancer [23]. Knowledge of their mechanisms of action may lead to new opportunities for preventing occupational cancer [23]. Mutations of the K-ras gene is frequent genetic changes of oncogenes in lung cancer [24]. Niklinska et al said that K-RAS oncogene exerts a crucial impact on cell cycle regulation and appears to be of major clinical significance for lung cancer evaluation [25]. Point mutations of the K-ras gene, which are found in 10 to 30% of lung adenocarcinomas, are regarded as being an early event during the carcinogenesis [26]. Autonomous vigorous motility of neoplastic cells, as well as growth and survival advantages, are considered to be necessary for cancer development and progression [26]. However, Ramakrishna et al reported that lung tumors did not have more total K-ras p21 or K-ras p21

GTP than normal lung tissue, nor were high levels of these proteins found in tumors with mutant K-ras [27]. They noted that normal K-ras p21 activity was associated with growth arrest of lung type II cells, and that the exact contribution of mutated K-ras p21 to tumor development remained to be discovered [27]. Okudela et al concluded that suggest that the K-ras-Akt pathway might facilitate the motility of neoplastic cells during the early period of carcinogenesis in lung adenocarcinomas, and might contribute to their non-invasive expansion along the alveolar septa, rather than invasion or metastasis [26].

In addition to lung cancer, the ras oncogene protein is also widely studied for its possibility in using as the tumor marker in pancreas and colorectal tumor marker. For pancreas cancer, genetic analysis of pancreatic juice is a promising aid for the accurate and early diagnosis of pancreatic cancer [28]. K- ras mutation is frequently observed in pancreatic cancer, however, it is not specific for carcinoma because pancreatic adenoma and pancreatitis also show this mutation [28]. Wenker et al also found that K-ras mutations lacked specificity to discriminate malignant pancreatic disease from chronic inflammation in tissue and stool [29]. For colorectal cancer, point mutations of the K-RAS gene at codon 12 are found in about 40% of cases with colorectal cancer, however, the diagnostic implications of the detection of these mutations and their clinical utility are still unclear [30]. Salbe et al noted that K-ras mutations could be found in circulating DNA extracted from serum samples of patients with colorectal cancer and showed that there was a correspondence between serum and tissue K-ras patterns [30]. It is noted that ras-p21 protein could be used as effective biomarkers for colorectal carcinogenesis in human chemopreventive trials [31].

De Biasi et al found that more ras p21 was observed in breast carcinomas compared with their respective normal counterparts [32]. It has been suggested that the immunocytochemical demonstration of the p21 ras oncogene product is a useful marker of malignancy in breast disease [33 – 34]. Rundle et al noted that ras p21 may be useful in the early detection of breast tumors and in post-surgical follow-up of patients, giving patients and physicians new tools for managing breast cancer [33]. However, Candlish et al proposed that the presence of ras p21 protein as demonstrated by the antibody technique was not a useful marker of malignancy or of proliferating epithelium but was rather a normal feature of certain cell types [34]. More ras p21 was also demonstrated in the majority of tumors of the stomach, colon and bladder compared with their respective adjacent normal tissues [32].

2. Fes Oncogene Protein

The human c-fes protooncogene encodes a protein-tyrosine kinase (c-fes) distinct from c-Src, c-Abl and other nonreceptor tyrosine kinases [35]. Although originally identified as the cellular homolog of several transforming retroviral oncoproteins, fes was later found to exhibit strong expression in myeloid hematopoietic cells and to play a direct role in their differentiation [35 - 36]. While it contains a tyrosine kinase and SH2 domain, there is no SH3 domain or carboxy terminal regulatory phosphotyrosine such as found in the Src family of kinases [36]. Fes has a unique N-terminal domain of over 400 amino acids of unknown function [36]. It has been implicated in signaling by a variety of hematopoietic growth factors, and is predominantly a nuclear protein [36].

The fes oncogene protein is also in the topic of occupational medicine at present, especially those who are at risk for leukemia [23].

3. Neu Oncogene Protein

The neu (c-erbB-2) oncogene and its p185 protein is mentioned as an important oncogene protein. The neu oncogene encodes a transmembrane phosphoglycoprotein [37 -39]. This molecule appears to be a growth factor receptor in the family of tyrosine kinase growth factor receptors [37 -39]. The neu gene encodes a novel growth factor receptor but, in contrast to met, its activation involves a single T:A→A:T point mutation in the region of the neu gene encoding the receptor transmembrane domain [37]. Makar et al suggested that the neu oncogene was an independent prognostic factor, which might predict the development of distant metastasis [38]. Preclinical studies have suggested that HER-2/neu overexpression enhances the metastatic potential of breast cancer cells [40]. Amplification and/or overexpression of c-erbB-2 in breast adenocarcinomas occurs frequently and its occurrence implies a more advanced malignancy [39]. This functional tumor marker is readily identified by appropriate DNA and antibody probes [39]. Breast carcinomas express high levels of ErbB receptors and their ligands, and their overexpression has been associated with a more aggressive clinical behavior [41]. Amplification of the HER-2/neu gene resulting in overexpression of the p185 occurs in approximately 25% of early stage breast cancers [42]. HER-2/neu has been established as an important independent prognostic factor in early stage breast cancer [42]. Overexpression /amplification of the HER-2/neu has been associated with a worse outcome in patients with breast cancer [42 – 45]. New data also suggest that HER-2/neu may be useful not only as a prognostic factor but also as a predictive marker for projecting response to chemotherapeutics, antiestrogens, and therapeutic anti-HER-2/neu monoclonal antibodies [42 - 44]. At present, monoclonal antibodies directed against HER-2/neu have been developed and used in clinical practice [40]. Clinical activity with one of these antibodies, trastuzumab, a humanized anti-ErbB2 MAb, has been documented in patients with breast cancer in a series of clinical trials and has recently been approved for the therapy of patients with metastatic ErbB2 overexpressing breast cancer [41]. Herceptin, a "humanized" murine monoclonal antibody directed against the extracellular domain of the HER-2/neu protein, is being used to treat breast cancer that overexpresses HER-2/neu [45]. The status of Her-2/neu in the tumor has become a critical factor in the management strategy of a breast cancer patient [45]. Therapeutic strategies that target other molecules involved in breast cancer development and progression are on the same parallel way [46]. In addition, in the breast cancer case, HER2/neu is a focus as a target for vaccine development [47]. It is crucial that pathologists become aware of these advances and assume a pivotal role in the development and application of assays to evaluate these new molecular targets [46].

4. myc Oncogene Protein

The myc proteins are nuclear proteins that exert their biological functions at least in part through the transcriptional regulation of large sets of target genes. Recent microarray analyses show that several percent of all genes may be directly regulated by myc [48]. The myc transcription factor functions as a downstream effector of most mitogenic signals [49]. Myc is synthesized rapidly in response to extracellular mitogenic signals, and blocking myc induction abolishes or at least severely attenuates any mitogenic response [49]. Furthermore, ectopic myc expression can often bypass the requirement for extracellular signals for entry into S phase [49]. It can be concluded that myc regulates to some degree every major process in the cell. Proliferation, growth, differentiation, apoptosis, and metabolism are all under Myc control and these processes feed back to adjust the level of c-myc expression [50]. Although myc is regulated at every level from RNA synthesis to protein degradation, c-myc transcription is particularly responsive to multiple diverse physiological and pathological signals [50]. However, it has been difficult to gain a precise understanding of how myc drives cancer because myc acts rather weakly at many of its target loci, and it has been reported to regulate as many as 10% to 15% of all cellular genes [51].

MYCN is a member of the myc family of oncogenes that encode nuclear proteins serving as transcription factors. This myc oncogene protein is mentioned for its tumor marker role for some cancers especially for neurological cancer. The correlation of N-myc gene amplification with poor prognosis in neuroblastoma was one of the first examples of prognostic data supplied by a c-oncogene [52].

In neuroblastoma, amplified MYCN is a strong prognostic indicator of poor prognosis [53]. Identification of amplified MYCN in neuroblastomas has marked the clinical debut of oncogenes, and MYCN status is now being widely used as a standard marker for neuroblastoma stratification [53 - 54].

5. P53

The tumor suppressor p53 is a transcription factor that is frequently inactivated in human tumors, therefore, restoring its function has been considered an attractive approach to restrain cancer [55]. Typically, p53-dependent growth arrest, senescence and apoptosis of tumor cells have been attributed to transcriptional activity of nuclear p53 [55]. Hence, p53 is not a pure oncogene protein tumor marker and will not be discussed in this chapter.

References

[1] Stone, MJ. Henry Bence Jones and his protein. *J. Med. Biogr*, 1998, Feb, 6(1), 53-7.

[2] Sproul, EE. Acid phosphatase and prostate cancer: historical overview. *Prostate.* 1980, 1(4), 411-3.

[3] Ochi, Y; Okabe, H; Inui, T; Yamashiro, K. Tumor marker--present and future. *Rinsho Byori*, 1997, Sep, 45(9), 875-83.

[4] Frydenberg, M; Wijesinha, S. Diagnosing prostate cancer - what GPs need to know. *Aust. Fam. Physician*, 2007, May, 36(5), 345-7.

[5] Devine, PL; McGuckin, MA; Ward, BG. Circulating mucins as tumor markers in ovarian cancer (review). *Anticancer Res.*, 1992, May-Jun, 12(3), 709-17.

[6] Fletcher, RH. Carcinoembryonic antigen. *Ann Intern Med*, 1986, Jan, 104(1), 66-73.

[7] Ishii, M. Limitation of clinical usefulness of tumor marker. *Gan. To Kagaku Ryoho*, 1995, Aug, 22(9), 1139-45.

[8] Motoo, Y; Watanabe, H; Sawabu, N. Sensitivity and specificity of tumor markers in cancer diagnosis. *Nippon Rinsho*, 1996, Jun, 54(6), 1587-91.

[9] Evans "It's a maybe test": men's experiences of prostate specific antigen testing in primary care. *Br. J. Gen. Pract.*, 2007, Apr, 57(537), 303-10.

[10] Marteau, TM; Kidd, J; Michie, S; Cook, R; Johnston, M; Shaw, RW. Anxiety, knowledge and satisfaction in women receiving false positive results on routine prenatal screening: a randomized controlled trial. *J. Psychosom. Obstet. Gynaecol*, 1993, Sep, 14(3), 185-96.

[11] Monier, R. Oncogenes and anti-oncogenes in tumorigenesis. *Reprod. Nutr. Dev.*, 1990, 30(3), 445-54.

[12] Della Porta, G; Radice, P; Pierotti, MA. Onco-suppressor genes in human cancer. *Tumori.* 1989, Aug 31, 75(4), 329-36.

[13] Levine, AJ. Tumor suppressor genes. Bioessays. 1990, Feb, 12(2), 60-6.

[14] Anderson, ML; Spandidos, DA. Onco-suppressor genes and their involvement in cancer (review). *Anticancer Res.* 1988, Sep-Oct, 8(5A), 873-9.

[15] Compagni, A; Christofori, G. Recent advances in research on multistage tumorigenesis. Br *J. Cancer*, 2000, Jul, 83(1), 1-5.

[16] Fotedar, R; Diederich, L; Fotedar, A. Apoptosis and the cell cycle. *Prog. Cell Cycle Res.*, 1996, 2, 147-63.

[17] Jeanteur, P; Theillet, C; Pujol, H. Oncogenes, anti-oncogenes and their alterations in human tumors. *Rev. Med. Interne*, 1990, May-Jun, 11(3), 216-20.

[18] Turcanu, V. An overview of oncogenesis. *Rev. Med. Chir. Soc. Med. Nat. Iasi*, 1992, 96 Suppl, 13-8.

[19] Renan, MJ. Cancer genes: current status, future prospects, and applications in radiotherapy/oncology. *Radiother .Oncol*, 1990, Nov, 19(3), 197-218.

[20] Leong, AS; Robbins, P; Spagnolo, DV. Tumor genes and their proteins in cytologic and surgical specimens: relevance and detection systems. *Diagn. Cytopathol*, 1995, Dec, 13(5), 411-22.

[21] McKenzie, SJ. Diagnostic utility of oncogenes and their products in human cancer. *Biochim. Biophys. Acta*, 1991, Dec 10, 1072(2-3), 193-214.

[22] Tong, LA; de Vos, AM; Milburn, MV; Jancarik, J; Noguchi, S; Nishimura, S; Miura, K; Ohtsuka, E; Kim, SH. Structural differences between a *ras* oncogene protein and the normal protein. *Nature*, 1989, Jan 5, 337(6202), 90-3.

[23] Brandt-Rauf, PW. Oncogenes and oncoproteins in occupational carcinogenesis. *Scand. J. Work Environ. Health*, 1992, 18 Suppl 1, 27-30.

[24] Mitsudomi, T; Takahashi, T. Genetic abnormalities in lung cancer and their prognostic implications. *Gan To Kagaku Ryoho*, 1996, Jul, 23(8), 990-6.

[25] Niklinska, W; Chyczewski, L; Niklinski, J. New molecular approaches to lung cancer: biological and clinical implications of P53, P16 and K-RAS studies. *Folia Histochem. Cytobiol*, 2001, 39(2), 99-103.

[26] Okudela, K; Hayashi, H; Ito, T; Yazawa, T; Suzuki, T; Nakane, Y; Sato, H; Ishi, H; KeQin, X; Masuda, A; Takahashi, T; Kitamura, H. K-ras gene mutation enhances motility of immortalized airway cells and lung adenocarcinoma cells via Akt activation:

possible contribution to non-invasive expansion of lung adenocarcinoma. *Am. J. Pathol.* 2004, Jan, 164(1), 91-100.

[27] Ramakrishna, G; Sithanandam, G; Cheng, RY; Fornwald, LW; Smith, GT; Diwan, BA; Anderson, LM K-ras p21 expression and activity in lung and lung tumors. *Exp. Lung Res.*, 2000, Dec, 26(8), 659-71.

[28] Mizumoto, K; Tanaka, M. Genetic diagnosis of pancreatic cancer. *J. Hepatobiliary Pancreat Surg.*, 2002, 9(1), 39-44.

[29] Wenger, FA; Zieren, J; Peter, FJ; Jacobi, CA; Muller, JM. K-ras mutations in tissue and stool samples from patients with pancreatic cancer and chronic pancreatitis. *Langenbecks Arch. Surg.*, 1999, Apr, 384(2), 181-6.

[30] Salbe, C; Trevisiol, C; Ferruzzi, E; Mancuso, T; Nascimbeni, R; Di Fabio, F; Salerni, B; Dittadi, R. Molecular detection of codon 12 K-RAS mutations in circulating DNA from serum of colorectal cancer patients. *Int .J. Biol. Markers*, 2000, Oct-Dec, 15(4), 300-7.

[31] Jia, X; Han, C. Biomarkers in the studies on chemoprevention of colorectal cancer. *Wei Sheng Yan Jiu.*, 2000, Mar 30, 29(2), 109-11.

[32] De Biasi, F; Del Sal, G; Hand, PH. Evidence of enhancement of the ras oncogene protein product (p21) in a spectrum of human tumors. *Int. J. Cancer.* 1989, Mar 15, 43(3), 431-5.

[33] Rundle, A; Tang, D; Brandt-Rauf, P; Zhou, J; Kelly, A; Schnabel, F; Perera, FP. Association between the ras p21 oncoprotein in blood samples and breast cancer. *Cancer Lett*, 2002, Nov 8, 185(1), 71-8.

[34] Candlish, W; Kerr, IB; Simpson, HW. Immunocytochemical demonstration and significance of p21 ras family oncogene product in benign and malignant breast disease. *J. Pathol.* 1986, Nov, 150(3), 163-7.

[35] Smithgall, TE; Rogers, JA; Peters, KL; Li, J; Briggs, SD; Lionberger, JM; Cheng, H; Shibata, A; Scholtz, B; Schreiner, S; Dunham, N. The c-Fes family of protein-tyrosine kinases. *Crit. Rev. Oncog*, 1998, 9(1), 43-62.

[36] Yates, KE; Gasson, JC. Role of c-Fes in normal and neoplastic hematopoiesis. *Stem .Cells*, 1996, Jan, 14(1), 117-23.

[37] Cooper, CS. The role of non-ras transforming genes in chemical carcinogenesis. *Environ. Health Perspect*, 1991, Jun, 93, 33-40.

[38] Makar, AP Neu (C-erbB-2) oncogene in breast cancer and its possible association with the risk of distant metastases. A retrospective study and review of literature. *Acta Oncol*, 1990, 29(7), 931-4.

[39] Maguire, HC; Jr; Greene, MI. Neu (c-erbB-2), a tumor marker in carcinoma of the female breast. *Pathobiology*, 1990, 58(6), 297-303.

[40] Sahin, AA. Biologic and clinical significance of HER-2/neu (cerbB-2) in breast cancer. *Adv. Anat. Pathol*, 2000, May, 7(3), 158-66.

[41] Albanell, J; Baselga, J. The ErbB receptors as targets for breast cancer therapy. *J. Mammary Gland Biol . Neoplasia*, 1999, Oct, 4(4), 337-51.

[42] Pegram, MD; Pauletti, G; Slamon, DJ. HER-2/neu as a predictive marker of response to breast cancer therapy. *Breast Cancer Res. Treat*, 1998, 52(1-3), 65-77.

[43] Stearns, V; Yamauchi, H; Hayes, DF. Circulating tumor markers in breast cancer: accepted utilities and novel prospects. *Breast Cancer Res Treat*, 1998, 52(1-3), 239-59.

[44] Wisecarver, JL. HER-2/neu testing comes of age. *Am. J. Clin. Pathol*, 1999, Mar, 111(3), 299-301.

[45] Kaptain, S; Tan, LK; Chen, B. Her-2/neu and breast cancer. *Diagn. Mol. Pathol*, 2001, Sep, 10(3), 139-52.

[46] Schnitt, SJ. Breast cancer in the 21st century: neu opportunities and neu challenges. *Mod. Pathol*, 2001, Mar, 14(3), 213-8.

[47] Wiwanitkit, V. Predicted B-cell epitopes of HER-2 oncoprotein by a bioinformatics method: a clue for breast cancer vaccine development. *Asian Pac. J. Cancer Prev.*, 2007, Jan-Mar, 8(1), 137-8.

[48] Kleine-Kohlbrecher, D; Adhikary, S; Eilers, M. Mechanisms of transcriptional repression by Myc. Curr. *Top Microbiol. Immunol*, 2006, 302, 51-62.

[49] Cole, MD; Nikiforov, MA. Transcriptional activation by the Myc oncoprotein. *Curr. Top. Microbiol. Immunol*, 2006, 302, 33-50.

[50] Liu, J; Levens, D. Making myc. *Curr Top Microbiol Immunol*, 2006, 302, 1-32.

[51] Knoepfler, PS. Myc goes global: new tricks for an old oncogene. *Cancer Res.*, 2007, Jun 1, 67(11), 5061-3.

[52] Duffy, MJ. Cellular oncogenes and suppressor genes as prognostic markers in cancer. *Clin. Biochem*, 1993, Dec, 26(6), 439-47.

[53] Schwab, M. MYCN in neuronal tumours. *Cancer Lett*, 2004, Feb 20, 204(2), 179-87.

[54] Phillips, WS; Stafford, PW; Duvol-Arnold, B; Ghosh, BC. Neuroblastoma and the clinical significance of N-myc oncogene amplification. *Surg. Gynecol. Obstet*, 1991, Jan, 172(1), 73-80.

[55] Fuster, JJ; Sanz-Gonzalez, SM; Moll, UM; Andres, V. Classic and novel roles of p53: prospects for anticancer therapy. *Trends Mol. Med*, 2007, May, 13(5), 192-9; Desmedt, EJ; De Potter, CR; Vanderheyden, JS; Schatteman, EA.

Index

A

B

C

F

G

H

I

J

K

L

M

N

O

P

Q

R

S

T

U

V

W

X

Y

Z